DISRUPTIVE

Rewriting the Rules of Physics

DISRUPTIVE

Rewriting the Rules of Physics

Steven B. Bryant

Infinite Circle
www.InfiniteCirclePublishing.com

Infinite Circle Publishing
3060 El Cerrito Plaza, Suite 364
El Cerrito, CA 94530
www.InfiniteCirclePublishing.com
Publisher@InfiniteCirclePublishing.com

Printed in the United States of America

ISBN 978–0–9962409–0–1 (Paperback)
ISBN 978–0–9962409–1–8 (Hardcover)
ISBN 978–0–9962409–2–5 (eBook)

Library of Congress Control Number: 2015913544

Publisher Reference Number: 06102016

Editing by Grant Dexter
Photography by Amy Slutak

10 9 8 7 6 5

Dedication

For my daughter Sarah, her generation, and all those who dare to ask: "Why?"

Table of Contents

Preface

"Why can't we go faster than the speed of light?" That was the question I asked everyone I knew. The answer came back: "Because Einstein said so," which was an answer that did not sit well with me. In fact, that specific answer pressed a particularly personal and emotional button that was implanted many years earlier. When I was a child, my mom would offer that response: "Because I said so!" when she was exasperated with me and wanted to bring the conversation to an end. It didn't mean she had an answer, at least not one that I would find remotely acceptable. However, it meant she was done discussing it and if I didn't want to get into trouble I had better "*step into line*."

I share this background to provide some context for the issue I had with people who gave it – "Because Einstein said so" – as a response to my question about *why* we could not travel faster than the speed of light. While this answer ended the conversation, it did not answer the underlying question: *Why?* As was the case with my mom's answer, their answer didn't sit well with me because it pressed the exact same button. So I had to find the answer another way. I had to discover the answer for myself.

The good news is that I knew that the answer was in Einstein's work. If people were saying: "Because Einstein said so," then I just needed to find *where* he said it. I knew that once I found it, I would either *step into line* and believe what everyone else believed, or I would have a better understanding of his work and find a specific reason for objecting and believing something different. In either case, I would have an answer. Finding the answer to this fundamental question is how I started my journey into the exciting world of physics.

While my exploration of physics started with a focused, concerted effort to understand Einstein's relativity theory, it did not end there. What was interesting is that for each question I answered, two new questions arose. It was this never–ending question–and–answer cycle that led to my understanding of the works by many of the founding fathers of theoretical and experimental physics: Newton, Maxwell, Lorentz, Einstein, Michelson and Morley, De Broglie, Heisenberg, and Planck.

When I started exploring physics, I approached my research with no preconceptions that any theory, specifically relativity theory, was right or wrong. My goal was to develop an understanding of the material based on what the authors said. I sought out their original works instead of looking for some simplification of their ideas that might be presented in an introductory college textbook. I wanted to understand the material so well that I could derive their equations and communicate their key theoretical points. I wanted to see everything I read from the point of view of the authors – to understand their choice of words (sometimes in their native languages) and understand why they chose certain words over other words. As an author, I know that how you say something – your choice of words; what you choose to say, and what you choose not to say – are all equally important. I think this perspective enhanced my understanding of the various theories, increased my understanding of what the authors were trying to accomplish, and helped me develop an understanding of what they were not able to effectively explain in their work.

I'm writing this book at a time when many scientists believe that certain theories, such as Einstein's relativity theory, are sacrosanct. As a result, we must acknowledge an important fact: Many of these theories have been useful for more than a century, with each helping us advance our scientific thinking and knowledge. However, just because a theory is useful, that doesn't mean it is perfect or, in some cases, even correct.

By today's standards, the ancient Greek model for the motion of the planets is wrong because it placed the Earth at the center of the solar system. Today, we know that all of the planets orbit the Sun as the center of the solar system, not the Earth. So, no matter how useful their model was and how well it might have served them, the Greek model was wrong. We have advanced beyond their model because our knowledge and experience have advanced. As a result, we are now able to make additional predictions that could not have been made with the ancient model.

Having a useful theory does not mean that we should end our quest for something better, something that has more accuracy, something that is easier to understand, or simply something that aligns better with other theories or experiments. The late Dr. Richard Feynman, a well–known and respected physicist, recognized the importance of being open to alternative theories, even those that disagree with the consensus. While the truth is probably in the prevailing direction, in his 1965 Nobel speech, he said that we have to remain open to the possibility that the "*truth may lie in another direction.*"

As responsible scientists, we must study, use, extend, and defend those models that provide the *best answers* and help us explain observations of our world. Now, we can't just change from something that seems to work to an alternative theory on a whim. That's not smart or useful. But, we always have to remain open to the possibility that each of our theories might be limited, incomplete, flawed, or simply wrong. If we're not open to this possibility, then we are not responsible scientists, because we will have let our beliefs get ahead of our judgment, rigor, and disciplined scientific method. So, we must explore those theories and ideas that give us the best answers – the best conceptual, mathematical, and predictive answers – regardless of our beliefs in the validity of any existing model or theory. This is how science advances.

This brings us to a very important point: A new model does not need to align with an existing theory, certainly not one that it is intended to replace. A new model must explain existing observations in a way that makes sense and is consistent. While we have to be able to explain existing theories and how they are related to a new theory, that comparison needs to happen at the end, not as a prerequisite.

We are going to develop and build our model from the ground up. As you progress through this book, you should gain a deeper understanding of this new model, called Modern Mechanics. Building upon classical mechanics as a foundation, Modern Mechanics is a unified model that offers greater mathematical accuracy than relativity theory and is easier to understand than quantum mechanics. In addition, you'll develop new perspectives of existing theories and experiments that you may not have seen before. For example, you'll learn about time dilation and length contraction, what they mean, why relativity theory requires both characteristics, and why these concepts are *not* part of Modern Mechanics.

You'll also examine nuances associated with one of the foundational experiments in quantum mechanics: the double–slit experiment. In this case, you'll learn how Modern Mechanics explains a particle's movement to the left or right of center in the experiment, rather than continuing in a straight line, a nuance not explained by quantum mechanics today.

This introduction to Modern Mechanics is intended for anyone with an interest in science, an interest in how things move, and an interest in physics. While an understanding of college algebra will facilitate comprehension, the math will be kept as simple as possible, only relying on advanced math when necessary. When we develop several equations in chapters 4 through 7, their meanings will be explained textually, as well as mathematically.

Sometimes, mathematics involving functions and equations can appear more difficult than it actually is. This can occur when we use an expanded character set, subscripts, or superscripts to represent variables. For example, if we wanted to sum the weight of Tom and Harry, we could represent this as the equation $S = T + H$. Alternatively, we could use Greek letters to write the equation as $\zeta = \delta + \varpi$. We could also use subscripted variables, where the same equation could be represented as $w_S = w_T + w_H$. While this may look complex and intimidating, we sometimes need more letters than the 26 that are in our alphabet. Greek letters, subscripts, or superscripts enable us to use other "letters" that make sense in our work. These are simply different ways of representing variables. For example, T, H, and S might refer to Tom's weight, Harry's weight, and their sum, as measured on Earth, while δ, ϖ, and ζ might refer to their weights as measured on Mars. Lastly, it is important to recognize that some characters, while different, will look similar. For example, the Greek symbol τ, called Tau, is different from the letter r, although they look similar.

I am excited to finally be completing this book. However, I don't want to give the impression that my work or thought process occurred in a vacuum. In fact, that would be entirely incorrect, especially given the number of people who have helped and inspired me. There are several people I'd like to acknowledge. I have to start with my wife, Julie, and daughter, Sarah. It is through them that I have found a never–ending source of love, passion, and support. I thank Irene Cole who, early on, advised me to approach my work in a scholarly, scientific way. That meant being disciplined about how I approached my research and writing. I also thank Dr. Nettie LaBelle–Hamer, whose advice after reading my first paper still rings loud and clear. She essentially said: *It's a good paper, but it doesn't go far enough. Don't just imply something; say it! People don't want to read something that isn't definitive. Put your neck on the table. Say what you want to say and stand by your convictions.* She was coaching me to take a risk, metaphorically, by putting my neck on the table. She was also advising me to make sure I had done my homework so that it didn't get chopped off. Great advice! It is advice that you will see reflected in the pages that follow.

I also want to thank my reviewers: Dr. Glenn Borchardt, Don Briddell, and Anatoliy Neymark. Each reviewer asked insightful questions, the answers to which are contained herein. Not only did they provide valuable feedback on the content, tone, and flow, their excitement and desire for the next update was truly inspiring. You will see their feedback reflected throughout the book.

Finally, I want to thank my editor, Grant Dexter. While he focused on grammatical, typographical, and content corrections, he read the material with the same passion and curiosity that I hope is shared by many readers. His attention to detail, corrections, and recommendations have enhanced and improved the overall quality. Words cannot express my gratitude for his contribution and I am fortunate to have partnered with such a skilled professional.

There are many others, who should be listed by name, but in more than 15 years of research it would take pages to list them all and I inevitably would leave someone out. At this point, I want to acknowledge everyone who has been part of my journey so far. Thank you!

I started writing this book in a café in Emeryville, California; with much of the material being written wherever I happened to be in the world: Berkeley, Charlotte, Point Reyes, London, San Francisco, or on an airplane flying at 36,000 feet. I hope that I'm able to provide you with an introduction to the new and exciting physics model called Modern Mechanics!

Steven Bryant, 2016
San Francisco Bay Area, California

Chapter 1 Introduction

Modern Mechanics is a unified model that represents the next generation in physics. It corrects mathematical mistakes and conceptual problems with existing theories, is intuitive and easy to understand, and has higher mathematical accuracy than its alternatives. To appreciate and understand what Modern Mechanics means to physics and why it's needed, we have to explain what physics is, understand the problems it tries to solve, and identify the problems with the established theories.

When people think of physics, what comes to mind for many is an extremely difficult subject that, if it wasn't part of a required science curriculum, might be actively avoided in high school and college. Others think of well–known theories like relativity or quantum mechanics, theories that some feel can only really be understood by a small subset of the scientific elite. These misperceptions about the level of difficulty of physics mask the simplicity and elegance that makes up its magnificent world.

Defining physics is surprisingly easy: *It is the study of why, how, and what happens when things move*. Consider several of the areas that make up physics and notice how each is simply a description of motion. Classical mechanics is about how things move around in the physical world. Thermodynamics, which is about heat, is in large part really about the movement of molecules, particles, or electrons to generate that heat. Optics is about how light behaves when it moves. Electrodynamics is essentially about how radio waves behave and move. Relativity is about the movement of large things and what happens when they move very fast. And quantum mechanics is about the movement of small things and what happens when they move very fast. If it moves, it is described with physics. This surprisingly simple explanation of what physics is helps us define a discipline that explains many aspects of our lives.

Since the beginning of time, mankind has tried to explain the world we live in, a large part of which involves describing how things move. Scientists play an important role by developing models and theories that explain what we are observing or provide us with a means to make predictions. Historically, these useful and valuable theories are grounded in religion and science. Sometimes explanations originate in religion and are later explained solely in scientific terms. Regardless of where an idea originates, as we expand our scientific knowledge and develop better models and theories, the number of things we are able to scientifically explain grows. Once a model or theory has demonstrated its value, it is difficult to change – even if it is not intuitive, has gaps in what it can explain, or has some inaccuracies or limitations. We would rather have a theory or model that is wrong and only works some of the time than nothing at all.

Replacing an existing theory or model is not an easy task, and acceptance of a new idea requires several elements; one of which is timing. It has to be developed at the right time and championed, meaning that it will be advocated by a person or group who will support and encourage its adoption over the reigning model. Generally, scientists are cautious about embracing a non–traditional, novel theory, especially if it negates an existing, well–established theory. This cautious approach minimizes the risk and harm to one's reputation that could result from challenging the accepted doctrine. Regardless of whether the new model is right, challenging the prevailing model is risky. Gaining the interest and support of the broad scientific community when they already support a position is a significant initial hurdle.

In addition, the new theory or model should be rational, intuitive, consistent, and offer something that the existing model does not. A model that does not offer better explanations, provide better answers, or make better predictions than the prevailing model is not going to make headway in advancing science or establishing itself as the leading model.

While successfully changing a widely accepted belief in an existing model or theory is difficult and does not occur often, it does happen. In fact, physics has already gone through two significant generational changes, or paradigm shifts. In the 15th century, the prevailing consensus was that the Earth was the center of the universe around which all heavenly bodies – the stars, planets, Sun, and Moon – orbited. Nicolaus Copernicus, someone whom today we would call an astronomer or a physicist, did not accept the reigning theory and had a different idea. He believed that the Sun was at the center of our solar system and that all of the planets orbited it instead of the Earth. In the 15th century, this was a radical idea, one that would take more than 200 years to be accepted. In fact, it did not change until the 17th century, when Galileo Galelei conducted several experiments that supported Copernicus' theory. Like Copernicus, Galileo's conclusions were considered radical; so much so that he was arrested, put on trial, and served the remainder of his life under house arrest. Considered heresy in his day, we now accept this Sun–centered solar system view as scientific truth. While change did not occur easily, the views of Copernicus and Galileo are now widely accepted, and their work on motion forms the foundation of physics and astronomy.

Isaac Newton, widely regarded as one of the greatest scientists who ever lived, was born less than a year after Galileo's death. Like Galileo, he conducted experiments in motion and developed several theories to explain how things move. The works of scientists like Copernicus, Galileo, and Newton, who are among the founding fathers of physics, form the foundation of first–generation physics, called classical mechanics.

Classical mechanics helps us understand how physical things move. Many students are introduced to classical mechanics in elementary school when they learn about geometric transformations in math class. Classical mechanics helps us answer questions about how we move throughout our world. For example, *if you live 300 miles from Disneyland and leave your house at 6AM, driving at 60 miles per hour, when will you arrive at Disneyland?* Many people quickly arrive at the answer of 11AM because it takes 5 hours to travel 300 miles when driving at 60*mph*. After adding 5 hours to a start time of 6AM, we know you will arrive at 11AM. Questions like this are answered on a daily basis and most of us never realize that we're using classical mechanics, or first–generation physics.

Classical mechanics served mankind well for a long time. It was, and still is, very useful in helping us understand how physical things move and behave. As the reigning theory of its day, it was beyond reproach until the discovery of a new and mysterious force called the electromagnetic force (EMF). While classical mechanics was extremely useful, it did not explain everything that scientists were observing with EMF experiments at the end of the 19th century. In fact, a scientific crisis occurred when experiments with light (also called optics) and EMF could not be explained using classical mechanics.

Out of crisis comes opportunity, and the turn of the 20th century was a fabulous time for physics. Our understanding of the electromagnetic force, the building block of all modern electronics and communications, was just forming. Because classical mechanics did not make accurate predictions in this area, we needed a new model or theory to explain how things moved. Scientists like Hendrik Lorentz, Henri Poincaré, and Albert Einstein filled the void by developing equations and theories to explain the experiments that could not otherwise be explained using classical mechanics.

Einstein's theory, called relativity, would reign for more than a century. Relativity theory is generally regarded as a replacement for classical mechanics and forms the foundation for modern theories in astronomy, motion, and gravitation. His theory, however, was not universally welcomed when it was first unveiled, probably because it was difficult to follow and because its conclusions marginalized the work of Newton, whose theories were still widely accepted. However, to Einstein's benefit, relativity theory provided an explanation for some of the experiments with light and EMF that classical mechanics was unable to explain. Because Einstein's theory offered an explanation, along with a mathematical model that offered far better accuracy than classical mechanics, it gained a solid foothold and became the foundational theory in what we now call modern physics.

Relativity theory was not the only theory that attempted to explain areas that fell through the gaps of what classical mechanics could explain. Led by scientists like Max Planck, Niels Bohr, and Werner Heisenberg, the field of quantum mechanics was born. While relativity explained the motion of large things, like planets, quantum mechanics explained the motion of very tiny things that are the building blocks of electrons and other small particles. Together, relativity and quantum mechanics are commonly referred to as modern physics, or second–generation physics.

While both theories do a good job of explaining their respective areas, neither serves as a unified theory, because they do not do a good job of explaining observations that the other explains well. First and second–generation theories and models were developed by different people at different times using different assumptions. As a result, these models and theories, and their accompanying mathematical equations, do not always work well with one another. A perfect example of this problem is the interplay between relativity theory and quantum mechanics, the latter of which relies heavily on statistics and probability. Einstein did not agree with quantum mechanics, saying it was "spooky" and that God does not roll dice. Because the two theories are based on different assumptions, their unification is challenging at best. While scientists today accept the merits of both theories (including their incompatibilities), we still love elegant solutions. *Why should we have two theories when one unified theory would be better?* Modern scientists have embarked down a path to find a theory that unifies relativity and quantum mechanics. This search for a unified theory is one of the most interesting and exciting scientific quests of the late 20th and early 21st centuries. Imagine the power, simplicity, and beauty of being able to explain with just one theory how things move and interact.

Previous attempts at a unified theory, like quantum electrodynamics and the standard model, retain and incorporate the existing theories. Any unified theory that tries to incorporate relativity as one of its foundational theories will be incorrect, since, as you will read later in this chapter, relativity is mathematically incorrect. This also applies to any derived theory, such as general relativity, which holds as one of its key assumptions that special relativity is correct. These are bold statements that can only be made because Modern Mechanics produces equal or better mathematical answers, in addition to overcoming mathematical and conceptual shortcomings found in earlier theories.

Modern Mechanics represents the third generation in physics and overcomes several problems with second–generation physics. First, quantum mechanics and relativity, while complementary, are at odds with one another in their fundamental assumptions. Modern Mechanics does not try to retain and reconcile the two. Instead it offers unified explanations of the key foundational experiments, independent of how other theories may have attempted to explain them in the past. Like Einstein's theory of relativity, it explains how large things move and behave. Like quantum mechanics, it explains how small things move and behave; it just does so differently. Second, some of the modern experiments with superluminal light (light traveling faster than 300,000km/s) are not easily explained by existing theories, or more worrisome, require experimenters to come up with novel reasons to explain why their results and the theory remain consistent. Such explanations are not required with Modern Mechanics, which does not have a *speed limit* like relativity theory. A third problem is that these modern–physics theories contain math or logical mistakes that invalidate them or, at a minimum, require their conceptual foundation to be revisited. These mistakes are extremely subtle, even by modern standards, and could not have easily been uncovered using the body of knowledge available to scientists at the end of the 19th or start of the 20th century.

Admittedly, this will be the most controversial aspect of this book, because many people will not easily accept that a theory that has been expertly reviewed and used for more than a century could contain an overlooked mistake. However, if we develop a new model that is more intuitive; has a unified explanation of the experiments addressed separately by classical mechanics, quantum mechanics, and relativity; and provides more accurate mathematical results than those models, then this would represent prima facie evidence of problems in those other theories.

Modern Mechanics has to satisfy three objectives to function as a unified theory. First, it must provide mathematical results that are equal to or better than the results provided by the reigning models. Any theory that fails to meet this objective cannot unseat the prevailing models. A significant benefit of producing equal or better math results is that scientists and engineers have a responsibility to research and understand those models and theories that offer the best mathematical answers. To do otherwise risks being mired in dogmatic belief, which is not scientific.

Second, it should be easy to understand. One of the challenges in modern physics is that relativity and quantum mechanics are hard for non–scientists to comprehend and understand. They are largely inaccessible to those who do not have a firm grounding in advanced math, including calculus and partial differential equations. While grounded in mathematics, Modern Mechanics is intuitive and can be largely described using geometric terms and algebra. Although any theory or model can be described using advanced math, and Modern Mechanics is no different, accessibility to the non–scientist is enhanced when we can communicate clearly and easily.

Third, it must be mathematically sound. This leads to two related questions. If modern physics has served us well for a century, what, if anything, could possibly be wrong with its foundational theories? Haven't they been "proven" by now? Surprisingly, wrong equations can often produce good results; up to a point. For example, as a crutch, some school children will learn that Pi equals 22 divided by 7. While incorrect, it is accurate to two decimal places, and if this is all the accuracy you need, this equation will work well. However, it is wrong because Pi, which begins as 3.14159265359, is defined as the circumference of a circle divided by its diameter. Pi, which is an extremely interesting irrational number, is not 22 divided by 7. Similarly, relativity contains mathematical mistakes that render it incorrect, but accurate up to a point. The small amount of error, when compared with other models of its day, explains why relativity theory has survived as a useful theory.

Relativity is one of the foundational theories in modern physics. It has produced good mathematical results and, in some cases, has been the only theory able to accurately predict the outcome of specific experiments. It has withstood the test of time by successfully repelling the previous onslaught of challenges and criticism. To understand why relativity has been so resilient and the mistake hasn't been previously discovered, we must examine how the theory has been challenged and defended.

As illustrated in Figure 1–1, relativity theory is actually the combination of two theories: special relativity, which Einstein defined in his famous 1905 paper, and general relativity, which Einstein published ten years later. General relativity is built upon the assertion that special relativity is correct, but incomplete. Both theories follow the same general approach, which is to 1) begin with a key set of *assumptions*, 2) develop the math for the theory – called the *derivation*, 3) check the work to ensure it makes sense and holds together – called the *proof*, and 4) discuss the key *implications* that result from the theory. Einstein's development of special and general relativity follow this four–step approach.

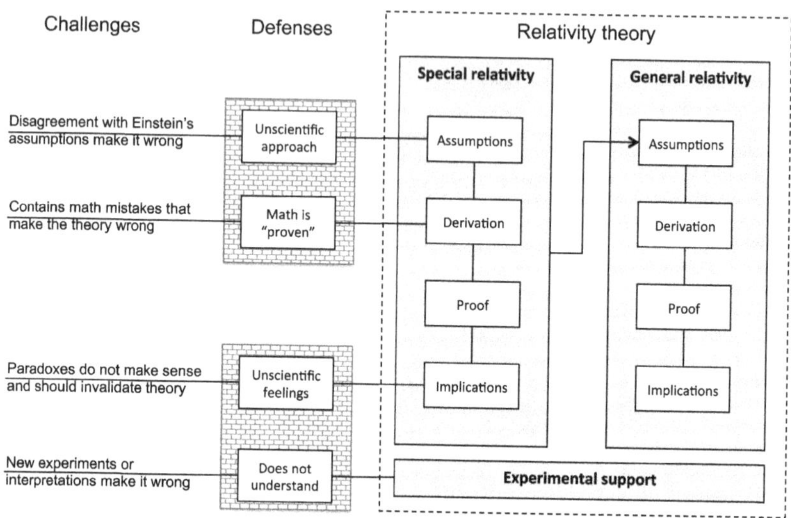

Figure 1–1 Framework for challenging and defending relativity theory.

Although Einstein's work follows this approach and is widely accepted, many have challenged his efforts. Most of the challenges have been made against special relativity, partly because it is the easier of the two theories to understand and partly because if special relativity were shown to be incorrect, then general relativity, because of its dependency on special relativity, would also be shown to be incorrect. Challengers have focused their attacks on special relativity's *assumptions*, *derivations*, *implications*, and *experiments*. Regardless of whether a challenge was on the right path or not, each of these attacks has been rejected.

Assumptions are simply the foundational items of a theory that are presumed to be correct. Scientifically, and in Einstein's work, these assumptions are called postulates. As long as the derivation, proof, and implications do not violate the foundational assumptions, then the theory holds together. Challenging a theory solely based on a disagreement over the foundational assumptions will be insufficient to invalidate a theory. All a supporter has to say is *"you don't understand the theory"* or *"you're not following a scientific process,"* dismiss the rest of the challenge, and walk away. This defense has proven successful because the challenger, especially when the rest of the theory remains unchallenged, is not following a disciplined mathematical or scientific approach.

Some challengers have attacked Einstein's *derivation*. This type of attack suggests that relativity theory contains mathematical mistakes that render the theory incorrect. However, there are two problems with this type of attack. First, the mistakes are extremely subtle and often hinge on an understanding of some specific nuances that not all scientists, engineers, and mathematicians are familiar with. However, what is often the case is that the challenger has overlooked a key mathematical characteristic. Second, the defense against this attack is to simply ask: *"If Einstein's math is wrong, why does his proof work and why do his equations offer the best math predictions?"* Until the introduction of *The Model of Complete and Incomplete Coordinate Systems*, the precursor to Modern Mechanics, this second objection could not be countered and the defense would hold. As you will soon read, the proof actually failed and the equations associated with Modern Mechanics produce equal or better results than relativity theory.

Rather than challenge the assumptions or derivation directly, some have resorted to challenging the *experimental* support for relativity theory. Often these attacks challenge the predictive capabilities of a theory without offering an alternative theory that would produce equal or better results. They interpret the results in novel, non–mathematical ways, or conduct new experiments that fail to gain the recognition of the broader scientific community. With this type of attack, the challenger often disregards the fact that relativity theory has a proven experimental track record. While this does not make the challenge wrong, these challengers fail because they do not provide an alternative that produces improved results, nor do they show what specifically is wrong with Einstein's theory. With this challenge, the defender simply says that the challenger "*doesn't understand the experiment*" and dismisses the attack.

Many challengers attack the *implications*, such as time dilation and length contraction, because these concepts do not make intuitive sense. They argue that the entire theory should be invalidated because the concepts of the theory do not make sense. The defense is simply to say, if the approach was sound; the assumptions are consistent; the derivation is correct; the proof passes; and the experiments support the theory, then you must accept the implications regardless of what you might intuitively believe. Unless the challenger can offer evidence of a problem elsewhere in Einstein's work, this type of attack will be easily defended because it is viewed as an undisciplined attack based on non–scientific *feelings*.

For more than a century, the nature of the challenges has fallen into these four categories: assumptions, derivation, experimental support, and implications. Since historically the challenges have been unsuccessful in convincing the broad scientific community of a problem with Einstein's work, many modern challenges in these areas are met with deaf ears.

The only area that had not been challenged was the *proof.* In fact, a quick review of Einstein's spherical wave proof would leave any reader with the belief that his proof is correct. Prior to the author's work, *The Failure of Einstein's Spherical Wave Proof*, few (if any) scientists had challenged Einstein's proof. It is straightforward to show where Einstein's proof fails, but it is far harder to change the *belief* that his theory is right. Before we can begin to define Modern Mechanics, it is imperative that you understand why people believe Einstein's proof is sound, why it actually failed, and why finding the mistake has been so elusive.

Einstein's Spherical Wave Proof

When scientists develop new, nonobvious theories, they must also develop proofs that show their assumptions and mathematics are compatible. Einstein developed such a theory and an accompanying proof, which is referred to throughout this book as the spherical wave proof. This proof is the most critical component of Einstein's paper, because without it he is unable to claim that his theory is mathematically and conceptually correct.

As illustrated in Figure 1–2, Einstein's spherical wave proof consists of six sentences. He begins with a spherical wave and considers the proof successful if he can show that the transformed points form a spherical wave when observed in a moving system.

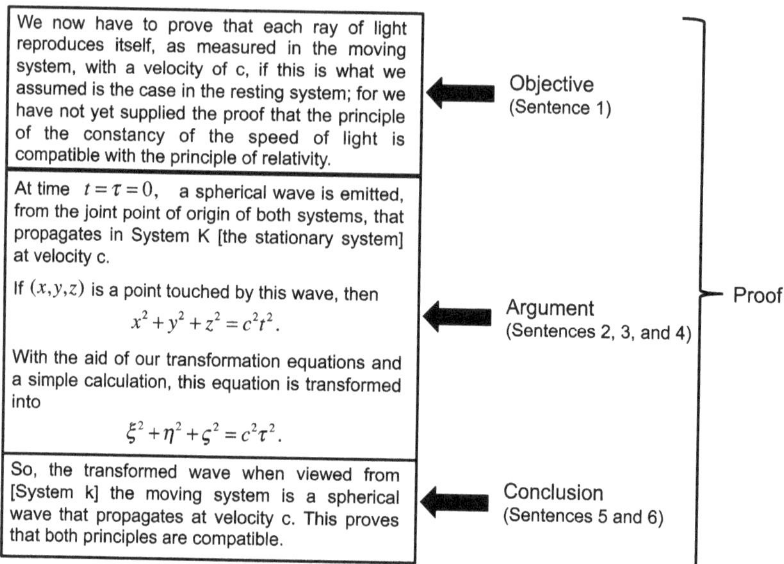

Figure 1–2 Spherical wave proof (English translation) from Einstein's 1905 paper establishing special relativity: *Zur Elektrodynamik bewegter Körper (On the Electrodynamics of Moving Bodies).*

Einstein's proof requires that light propagate in all directions – up, down, left, right – at the same rate, regardless of whether the system, also called a *frame*, is in motion or stationary (Sentence 1). A spherical wave is a conceptualized spherical surface whose size expands in all directions at the same rate. A way to visualize a spherical wave is to imagine a large bubble that gets bigger as more air is blown into it. A spherical wave, which can be thought of as a “light bubble,” expands in all directions at the same velocity, c. Regardless of how big the wave becomes, it remains spherical.

The remainder of the proof is straightforward:

1. Begin with a specific closed shape, a spherical wave, which is just an expanding wave (or bubble) whose radius from a common center is determined by the amount of time that has passed since the wave was first emitted (Sentence 2). Einstein chose a spherical wave because all of its light rays extend outward from the center at the same velocity. In addition, a spherical wave enables Einstein to show that his theory is valid for *every point in three–dimensional space*; a requirement for his proof to be valid.

2. Take all of the points that make up the surface of the spherical wave and confirm that they satisfy the mathematical equation for a sphere:

$$x^2 + y^2 + z^2 = c^2t^2 \qquad \text{Eq. 1.1}$$

   where ct is the radius (Sentence 3). Einstein uses ct rather than R in his equation to represent the radius from a common center in the stationary frame because he is discussing a spherical wave, which has a specific size at time t. This is distinguished from a static sphere, whose size does not vary with time. Einstein uses the Latin characters x, y, and z, to represent a position, and t to represent time in the stationary frame.

3. Take all of the points that make up the first sphere and use his equations to produce "transformed" points (Sentence 4).

4. Take the "transformed" points and show that they satisfy the mathematical equation for a sphere:

$$\xi^2 + \eta^2 + \varsigma^2 = c^2\tau^2 \quad \text{Eq. 1.2}$$

   where $c\tau$ is the radius (Sentence 4). Similar to what was described in Step 2, Einstein uses $c\tau$ rather than R' in his equation to represent the radius in the moving frame. This is distinguished from a static sphere whose size does not vary with time. Einstein uses the Greek characters ξ, η, and ς, to represent a position and τ to represent time in the moving frame.

5. Conclude that if steps 1 through 4 are successfully performed, a spherical wave is formed in the moving frame, successfully completing the proof (sentences 5 and 6). If a spherical wave is not formed in the moving frame, then the proof establishing relativity theory fails.

As illustrated in Figure 1–3, Einstein and other researchers over the past century have reviewed these five steps to reach the same conclusion; that a spherical wave was formed in the moving frame and the proof appears to pass.

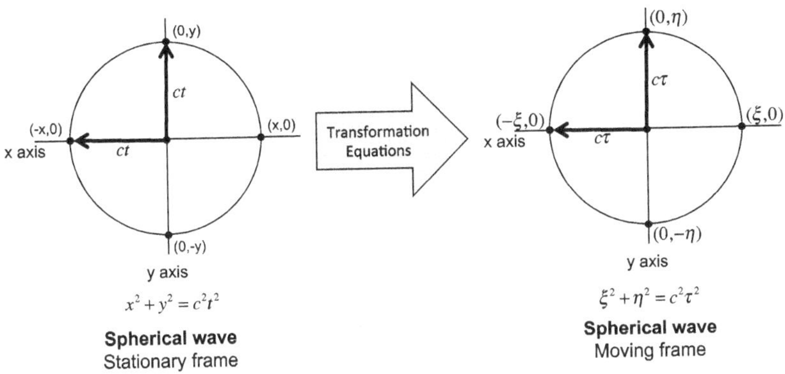

Figure 1–3 The purpose of Einstein's spherical wave proof is to show that if a spherical wave exists in the stationary frame, the converted points will all be part of a transformed spherical wave in the moving frame. (Note: A two–dimensional circle is shown for illustrative purposes.)

While not obvious, the proof actually failed, because a spherical wave is not formed in the moving frame. The problem is hard to detect, because if you follow steps 1 through 5, above, you will reach the same conclusion as Einstein. The problem with Einstein's proof is not that the steps are wrong; the problem is that the proof is incomplete. As you will soon see, satisfying the equations is alone insufficient to prove that a spherical wave exists in the moving frame. To understand why this proof failed, we have to examine the definition of a sphere.

> *A **sphere** is a three–dimensional surface, **all points of which** are equidistant from a fixed point.* *Def. 1.1*

It follows that:

> *A **spherical wave** is a conceptual three–dimensional surface, **all points of which** are equidistant from a fixed point, where the radius is determined by the velocity of the*

wave and the amount of time since the wave was first emitted. *Def. 1.2*

A sphere and a spherical wave share an important characteristic: that *all points are equidistant from a common fixed point.* The transformed shape must satisfy this definition for Einstein's objective and conclusion to be satisfied. If the second shape meets the definition, the proof passes; if it does not, the proof fails.

Before examining Einstein's proof further, we must look at the conditions required to determine whether a set of points satisfies Definition 1.1 and by extension Definition 1.2. In mathematics, the condition: "***all points of which*** *are equidistant from a fixed point*" is often implied or assumed. It is not explicitly checked. For example, a two–dimensional surface, called a circle, is defined as the set of all points equidistant from a common fixed point. It is mathematically written as:

$$x^2 + y^2 = R^2 \qquad \text{Eq. 1.3}$$

The points on the surface of the circle are written as members of a set in the form $\{(x, y)\}$. To simplify our analysis, we will only examine four specific points, which fall on the x or y axis:

$$\begin{Bmatrix} (-1,0), \\ (0,1), \\ (1,0), \\ (0,-1) \end{Bmatrix}$$

Every point is assumed to be the same distance from the center (0,0). When given this set of points, we would use Equation 1.3 to correctly conclude that they are part of the same circle. Although not explicitly checked, the radius of 1 is assumed true for every point that lies on the circle.

Alternatively, the radius can be explicitly included as a member of the set, so that the points are written in the form $\{(x, y, R)\}$. Using the same example as given above, we would simply include the radius with each coordinate, so that the four points are now written as:

$$\begin{Bmatrix} (-1,0,1), \\ (0,1,1), \\ (1,0,1), \\ (0,-1,1) \end{Bmatrix}$$

As illustrated in Figure 1–4, when this set of x, y, and R values is validated using Equation 1.3, we once again correctly conclude that they are all part of the same circle.

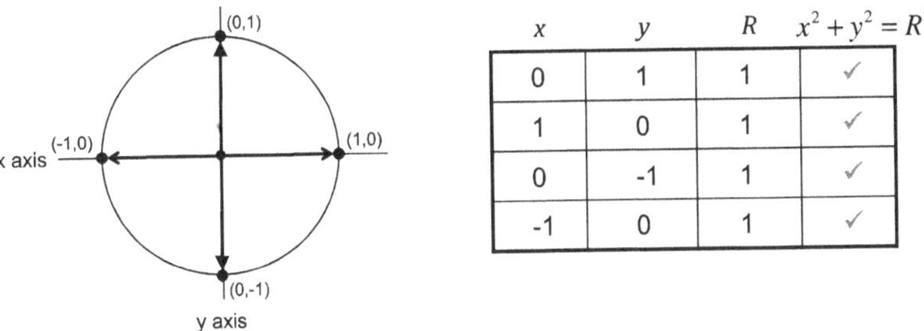

x	y	R	$x^2 + y^2 = R^2$
0	1	1	✓
1	0	1	✓
0	-1	1	✓
-1	0	1	✓

Figure 1–4 The set of values that all satisfy the equation of a circle (Equation 1.3) with all points having the same radius.

The inclusion of the radius as part of the set is subtle, but critically important to understanding the problem in Einstein's proof. To illustrate the effect of this change on the behavior of the equation, consider the following set of four points, also written in $\{(x, y, R)\}$ notation:

$$\begin{Bmatrix} (1,1,1.41), \\ (2,2,2.83), \\ (3,3,4.24), \\ (4,4,5.66) \end{Bmatrix}$$

Each point is expressed in terms of x, y, and the radius R (rounded to two significant figures). Since Equation 1.3 is satisfied for each point in the set, we could incorrectly conclude that these points are all part of the same circle. This is illustrated in Figure 1–5, where you can see that the points form a line, not a circle. What this means is that the use of the equation alone, without validating that all points share the same radius, will not prove that the points are part of any specific shape. *With the radius included with the coordinates for a point, we must show that the radius is the same for every point in the set.*

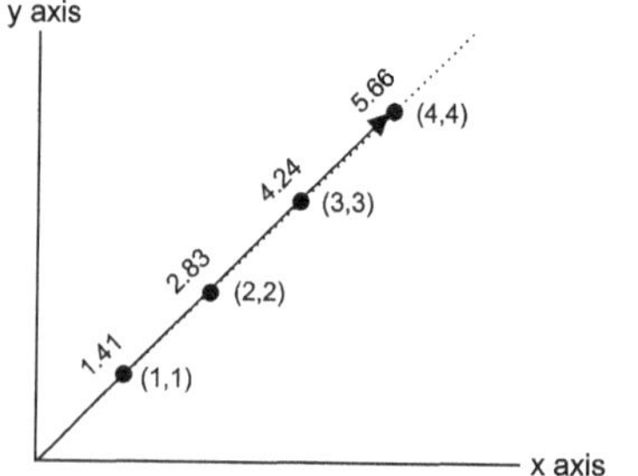

x	y	R	$x^2+y^2=R^2$
1	1	1.41	✓
2	2	2.83	✓
3	3	4.24	✓
4	4	5.66	✓

Figure 1–5 A set of values that individually satisfy the equation of a circle (Equation 1.3), leading to the incorrect conclusion that they are all points on the same circle. These points do not form a circle, since they do not all share the same radius.

Equation 1.3 will correctly tell us when the members of a set are part of a valid circle. Unfortunately, Equation 1.3 could also tell us that we have a valid circle when, in fact, we do not. This type of mistake is called a Type I error and occurs when correct results are returned (eg, a circle) along with results that should have been rejected (eg, a line).

Returning to Einstein's proof: To show that a spherical wave was formed in the second frame, definitions 1.1 and 1.2 must be satisfied. This means we must show that the set of values each satisfy the equation of a sphere *and that all of the values share the same radius.* This requires us to perform a second test to confirm that there is only one unique radius in the set of values.

Mathematically, because we define the set A to contain the distance from the origin for each point on the shape, the second check must show that $|A| = 1$. While the cardinality operator $|A|$ looks identical to the absolute value operator, it is used to count the number of unique values in a set. For example, consider the set $A = \{1,1,1,1\}$, which represents the radii for the points illustrated in Figure 1–4. Because there is only one unique distance in the set, its cardinality is $|A| = 1$. In other words, every point is the same distance from the origin. However, when the set is $A = \{1.41, 2.83, 4.24, 5.66\}$, which represents the distances from the origin for the points illustrated in Figure 1–5, then $|A| = 4$. Because the cardinality of this set is greater than 1, all points do not share the same radius. A circular or spherical surface requires that the distances from a common center to each point be the same, requiring $|A| = 1$. This cardinality check will confirm that every point is the same distance from the center.

We now examine the five steps of Einstein's spherical wave proof using a brute–force analysis, which uses specific values. Step 1 requires that we begin with a valid spherical wave in the stationary frame. This condition is met if we evaluate the spherical wave at $299{,}792{,}458^{-1}$ seconds following emission. At this time, the wave has a radius of 1 meter in all directions. This is called a unit sphere and we will examine four points on its conceptual surface:

$$\begin{Bmatrix} (-1,0,0,\frac{1}{299{,}792{,}458}), \\ (0,1,0,\frac{1}{299{,}792{,}458}), \\ (1,0,0,\frac{1}{299{,}792{,}458}), \\ (0,-1,0,\frac{1}{299{,}792{,}458}) \end{Bmatrix}$$

The set of values includes time, t, and is written in the form $\{(x,y,z,t)\}$, with z required for calculation of a sphere as opposed to only x and y required for a circle in previous discussions. The radius is easily found by multiplying time, t, by the velocity of light, c, for each point in the set. Given that $c = 299{,}792{,}458m/s$, it is easy to show that each point has the same radius, which is 1 meter.

Step 2 uses Equation 1.1 to successfully confirm that all of the points of the original unit sphere actually form a spherical surface. This condition is satisfied.

Step 3 converts the original points $\{(x,y,z,t)\}$ into a set of transformed points $\{(\xi,\eta,\varsigma,\tau)\}$. All of the points of the original spherical wave are converted using Einstein's equations:

$$\xi = \frac{x - vt}{\sqrt{1 - \frac{v^2}{c^2}}}$$

$$\eta = y$$

$$\varsigma = z$$

$$\tau = \frac{t - \frac{vx}{c^2}}{\sqrt{1 - \frac{v^2}{c^2}}}$$

To visually perform the analysis, we select a large velocity, $v = 289{,}000{,}000 m/s$, so that the first point

$$(-1,0,0,\frac{1}{299{,}792{,}458})$$

is transformed into:

$$(-7.4,0,0,\frac{7.4}{299{,}792{,}458})$$

Each point is converted into a transformed point using Einstein's equations, the results of which are given in Table 1–1.

	Original Points				Transformed Points			
Row	x	y	z	t	ξ	η	ς	τ
1	1	0	0	3.3356E-09	0.1	0	0	4.5160E-10
2	-1	0	0	3.3356E-09	-7.4	0	0	2.4638E-08
3	0	1	0	3.3356E-09	-3.6	1	0	1.2545E-08
4	0	-1	0	3.3356E-09	-3.6	-1	0	1.2545E-08
5	0	0	1	3.3356E-09	-3.6	0	1	1.2545E-08
6	0	0	-1	3.3356E-09	-3.6	0	-1	1.2545E-08

c= 299,792,458
v= 289,000,000

Table 1–1 Conversion of the points on the unit sphere into transformed points using Einstein's equations. Position coordinates are written to one significant figure. Time values are in scientific notation. For example, $3.3356\text{E} - 09$ is written in fractional notation as $299{,}792{,}458^{-1}$ in the text. Time is measured in seconds and velocity is measured in meters per second. Points that fall on the z axis (rows 5 and 6) are shown for completeness, but not discussed in the text.

Once the original points are converted using Einstein's equations, the transformed points are:

$$\left\{\begin{array}{l}(-7.4,0,0,\frac{7.4}{299{,}792{,}458}),\\(-3.6,1,0,\frac{3.8}{299{,}792{,}458}),\\(0.1,0,0,\frac{0.1}{299{,}792{,}458}),\\(-3.6,-1,0,\frac{3.8}{299{,}792{,}458})\end{array}\right\}$$

The radius for each transformed point is found by multiplying time, τ, by the velocity of light, c. Because relativity requires that $c = 299{,}792{,}458 m/s$ for each of the transformed points, it is easy to see that the length of each ray that begins at the origin of the new shape is found using $R' = c\tau$ for each of the converted points. To make the radius obvious, the points are given in the form $\{(\xi, \eta, \varsigma, R')\}$ as:

$$\left\{\begin{array}{l}(-7.4,0,0,7.4),\\(-3.6,1,0,3.8),\\(0.1,0,0,0.1),\\(-3.6,-1,0,3.8)\end{array}\right\}$$

Step 4 confirms that all of the transformed points, to one significant figure, satisfy Equation 1.2. Since the equality of this statement is maintained for each point, this condition is satisfied.

Since the requirements of steps 1 through 4 of the proof were successfully met, we would conclude in Step 5 that a valid spherical wave was formed in the second frame. In other words, we have just shown that Einstein's proof appears to work. However, as illustrated in Figure 1–6, the transformed shape is not a spherical wave centered at the origin, but is instead an ellipsoidal wave with a center to the left of the origin at $(\frac{-vt}{\sqrt{1-\frac{v^2}{c^2}}}, 0, 0)$.

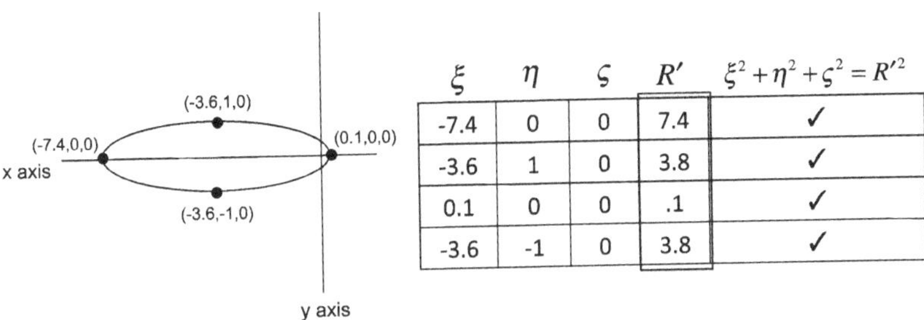

ξ	η	ς	R'	$\xi^2+\eta^2+\varsigma^2=R'^2$
-7.4	0	0	7.4	✓
-3.6	1	0	3.8	✓
0.1	0	0	.1	✓
-3.6	-1	0	3.8	✓

Figure 1–6 The set of transformed points individually satisfy the equation of a sphere, leading to the premature conclusion that they are all points on the same spherical wave. This is a Type I error. The transformed shape is not a spherical surface centered at the origin, but is instead an ellipsoidal surface with a center to the left of the origin. *Note: The transformed values are rounded, which does not change the diagram or conclusions.*

An ellipse is not a sphere. In an ellipse, the radii (the distances from the center to the points on the surface) are not the same for all points. Mathematically, an ellipsoidal surface is described in terms of its center point and the length of its semi–minor and semi–major axes. Positions on the surface of the ellipsoidal wave are determined by the position of its foci on the major axis and by the length of the cord connecting the foci. While the stretching effect of the ellipsoidal surface along the x axis and shift of the center to the left of the origin is most *visibly* pronounced at high velocities, it is mathematically true for all positive velocities regardless of magnitude.

Now we revisit definitions 1.1 and 1.2 to determine whether the transformed shape satisfies these definitions. The transformed shape does not satisfy these definitions because *all points are* ***not*** *equidistant from a common fixed point*. In other words, because $|A| = 1$ is not true, the transformed shape does not meet the definition for a sphere or spherical wave.

As illustrated in Figure 1–7, there are actually four reasons that the transformed shape does not satisfy the definition and the proof fails. First, the two shapes are not the same; one is spherical while the other is ellipsoidal. In the spherical wave, the semi–major and semi–minor axes are the same length. In the ellipsoidal wave, the semi–major and semi–minor axes have different lengths. Second, in the spherical wave, the origin is at the center, while in the ellipsoidal wave the origin is to the right of center. Third, in the spherical wave all points share the same length (or radius) from a fixed point. However, in the ellipsoidal wave, the distances to the points from a common fixed point are different. Fourth, in the spherical wave, the distances, ct, are radial values, where each represents a distance from the center of the shape to its surface, but in the ellipsoidal wave, the distances, $c\tau$, represent a type of generator line that represents the distance from the origin (not the center of the shape) to points on the surface of the ellipsoid.

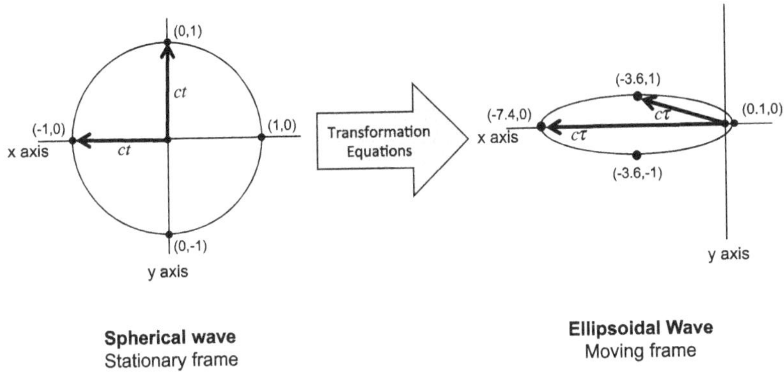

Figure 1–7 Einstein's equations transform a spherical wave into an ellipsoidal wave. In the original shape, illustrated by the left–hand image, the radius to reach each point is the same, originates from the center of the shape, and travels the same amount of time, t. Points on the transformed shape have different values for time, τ, and as a result have different lengths from the origin. In the transformed shape, illustrated in the right–hand image, the origin is not the center of the shape.

As noted earlier, this is a Type I error. It is very hard to detect because it requires the explicit examination of the radius for every point in the transformed shape. When all or part of the radius is included as part of a coordinate for a point, Equation 1.2 can only be used as a distance equation. However, since values are provided on both sides of the equals sign, the equation cannot distinguish between a collection of points belonging to a spherical wave, an ellipsoidal wave, a straight line, or a random collection of points. This is summarized in Figure 1–8.

	Original spherical wave	Transformed ellipsoidal wave	Collection of points on a line	Random collection of points
All points satisfy the equation when values from both sides of the equals sign are used	✔	✔	✔	✔
All points share the same radius	✔	✖	✖	✖
Origin is the center of the shape	✔	✖	N/A	N/A
Semi-major and semi-minor axes are the same length	✔	✖	N/A	N/A
Velocity multiplied by time represents the distance to a point from the center of the shape	✔	✖	N/A	N/A
Shape is spherical	✔	✖	✖	✖

Figure 1–8 When a constant radius is not common for all points, equations 1.1 and 1.2 cannot be used to confirm the existence of any specific shape. The additional check for a constant radius is necessary to confirm a spherical shape.

Einstein thought he had a spherical wave when, in actuality, he did not. A spherical wave requires a constant radius such that the same $R' = c\tau$ is true for all points. Said mathematically, $|A| = 1$ must be true for the transformed wave. Since τ has different values for the transformed points, the only way to maintain a constant radius, R', in the transformed frame is for the velocity of light, c, to change. Specifically, c would need to change with each different x coordinate in the transformed points. This would make Einstein's conclusion that "*the transformed wave when viewed from [System k] the moving system is a spherical wave that propagates at velocity c*" false, because c could not be constant: It must change in the moving frame to maintain a constant radius, R', for all points. If the velocity for each ray changes, then not only would this make Einstein's conclusion false, the equality of the mathematical equation would also be false. As a result, relativity theory is not validated, since Einstein's proof failed to associate the *principle of the constant velocity of light* with the *principle of relativity*.

The failure of the proof lends support to many of the historical challenges made against Einstein's work. A key strength of this analysis is that it does not require the introduction of new or novel mathematical concepts. Rather, the reader is simply reminded of the definition of a sphere (Definition 1.1) and is asked to verify whether that condition was satisfied. Another strength of this analysis is that, had a valid spherical wave been formed, the additional check for a constant radius would not change the proof's result. The fact that a check for a constant radius alters the proof's conclusion, changing it from pass to fail, supports the finding of a Type I error.

Common Objections

Even in the face of this evidence, some of Einstein's defenders will assert that the proof works because Equation 1.2 is satisfied and that the points do not need to form a spherical wave. This defense fails because it requires you to ignore half of the proof: the objective and the conclusion. Einstein is unambiguous in both his objective and his conclusion, where his intent was to show that if you begin with a spherical wave, you must end with a spherical wave.

Other defenders will argue that Einstein is only talking about one specific point on the sphere because, in some English–language translations, the first sentence reads "*any* ray of light," not "*each* ray of light." First, the use of "each" or "every" is a more accurate reflection of the original German paper. Second, once again, this defense fails because it also ignores half of Einstein's proof. In fact, this second defense is simply a variation of the first. Both defenses show a lack of understanding of Einstein's work, since satisfying the equations alone, or just considering a single point, would not prove the relationship between Einstein's two postulates.

A third defense asserts that a spherical shape has been formed if you *assume* simultaneity, length contraction, time dilation, or some combination of those terms. This defense fails because it does not recognize that the transformed points found using the transformation equations belong to the moving frame. The defender doesn't care what the math says and suggests that the proof should just be taken as valid because of a term Einstein invented. As a result, this defense requires you to disregard the proof's mathematical argument, specifically sentences 4 through 6. It also fails because, in a logical argument, a conclusion resulting from a proof – or the use of the term simultaneity, time dilation, or length contraction – cannot be used in defense of that proof *prior to the proof's completion*. While we will examine these relativistic terms in Chapter 6, what is important to know is that one cannot use any of Einstein's relativistic terms as a defense until after the proof establishing relativity has been successfully completed.

Each defense ignores the purpose of the spherical wave proof and, at a minimum, requires you to ignore half of Einstein's proof: the objective and the conclusion. Additionally, the terminology defense requires you to ignore the proof's mathematical argument as well. Said simply, the terminology defense ignores the entire proof.

The failure of Einstein's spherical wave proof does not indicate exactly what is wrong with Einstein's work. It just shows that there is a problem that results in the proof's failure. To determine why the proof's failure will ultimately invalidate relativity theory, we must first identify what is wrong in Einstein's work. This will be examined in Chapter 6 when we re–examine Einstein's 1905 derivation, equations, and terms.

The Michelson–Morley Experiment

The failure of the spherical wave proof will not be sufficient for some to accept a problem in Einstein's work. They will continue to believe that "*Einstein's work is experimentally proven.*" To dispel this belief, we have to revisit a foundational experiment long thought to support relativity theory: the 1887 Michelson–Morley interferometer experiment. Albert Michelson and Edward Morley, two early researchers in electromagnetic force and optics, performed an experiment to validate a classical mechanics–based theory proposed by French engineer and physicist Augustin–Jean Fresnel. They hypothesized that if light were traveling as a wave, experiments would detect its motion, which they could use to determine the Earth's velocity as it orbited the Sun. They built an ingenious device called an interferometer, which they expected would detect the wave motion of light, allowing them to calculate the Earth's orbital velocity, which they knew was about 30km/s. Said simply, they built a speedometer and used it to measure how fast the Earth was traveling around the Sun.

As illustrated in Figure 1–9, after collecting and analyzing their data, Michelson and Morley were only able to **compute** an Earth orbital velocity of about 8km/s. This was a far cry from the 30km/s they expected to detect. It didn't work. Their experiment did not support the wave–based theory of light suggested by Fresnel. Einstein's supporters have taken this failure and turned it into a success for relativity theory. Supporters say that the Michelson and Morley measurements are anomalies, or errors, inherent with the interferometer. Their argument is that the device detected a velocity of 0km/s, which supports relativity theory, and that the observed measurement is simply "experimental error." This experiment and their revised conclusion of 0km/s is one of the reasons people believe that relativity theory has been experimentally "*proven.*"

Fortunately, scientists and mathematicians have a disciplined way of explaining "experimental error" called *statistics*. Because Michelson and Morley included their data in their paper, we are able to perform a statistical analysis of their results. In statistics, a confidence interval defines the range in which we would expect to find the actual result, assuming a specific probability. For the Michelson–Morley experiment, this range is represented by the "normal curve" in Figure 1–9. What this diagram says is that we are 99.9% sure that the Michelson–Morley experimental data, *when evaluated using their equations*, found the Earth orbital velocity to be between 6km/s and 10km/s. By expressing the answer in statistical terms, we account for any errors that might exist with the experimental device or measurements. Clearly, 30km/s does not fall within the range, prompting Michelson and Morley to conclude that their results did not support Fresnel's theory. This led to a crisis, because classical mechanics was unable to predict these results, providing the opening for relativity to provide a better answer and establish itself as a leading theory.

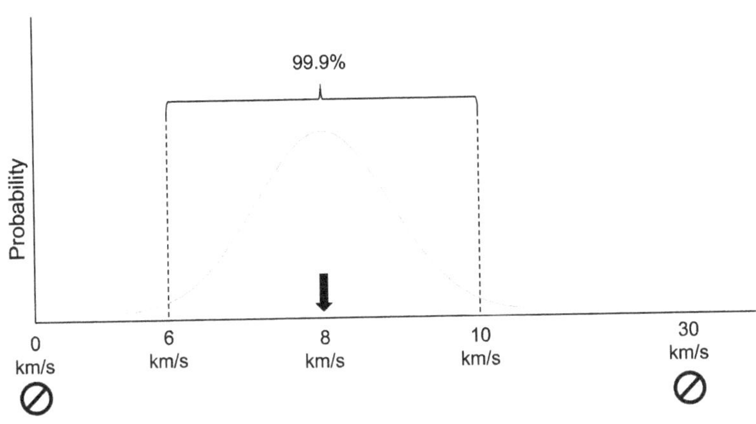

Figure 1–9 Results of the original Michelson–Morley interferometer experiment illustrated as a 99.9% confidence interval for the experiment. The amount of error between the actual result of 8km/s and the expected result of 30km/s is 22km/s. The amount of error between the actual result of 8km/s and the relativistic expected result of 0km/s is 8km/s. Neither 0km/s nor 30km/s fall within the 99.9% confidence level, suggesting that the Michelson–Morley analysis and data do not support Fresnel's theory or relativity theory. Illustration is not to scale.

However, 0km/s, which is required by relativity theory, does not fall within this range either! Mathematically, we are less than 0.1% confident that 0km/s is the Earth orbital velocity. The Michelson–Morley experiment does not support relativity theory when viewed through a statistical lens. Any theory based on the Michelson–Morley experiment returning 0km/s as the answer has less than a 0.1% chance of being right. In other words, the only way the Michelson–Morley experiment supports 0km/s, as required by relativity theory, is if we completely ignore the experiment's actual result!

Furthermore, if we do not ignore the actual data and do the calculations properly, we obtain a value supporting Fresnel's original hypothesis. As illustrated in Figure 1–10, and in contrast to the relativistic interpretation, the amount of error (the difference between the actual result and the expected result) when their data is evaluated using Modern Mechanics is less than 3km/s for the Michelson–Morley experiment. In fact, the error is less than 0.3km/s with Dayton Miller's repeat experiment performed in 1933. In both experiments, the error is significantly less when the Modern Mechanics equations are used to analyze the data. More importantly, the expected result of 30km/s is extremely close to the properly computed result of 32km/s. The expected result falls within the 99% confidence interval, showing that their experiment worked!

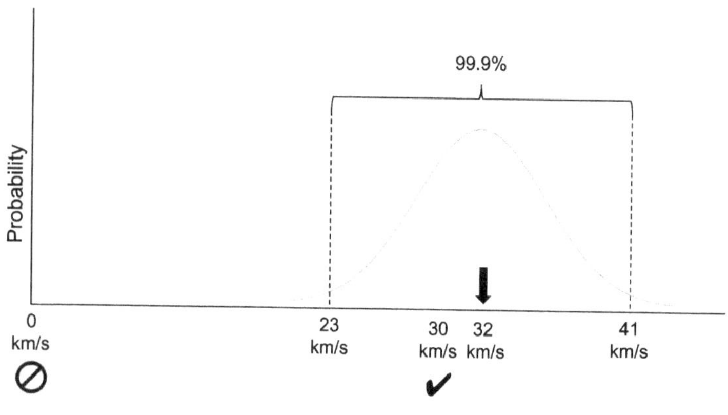

Figure 1–10 Results of the Michelson–Morley interferometer experiment evaluated using Modern Mechanics' equations. The amount of error between the actual result of 32km/s and the expected result of 30km/s is 2km/s. In addition, the expected result of 30km/s falls within the 99.9% confidence interval, suggesting that the Michelson–Morley data when analyzed using Modern Mechanics' equations, supports the result. Illustration is not to scale.

The Michelson–Morley experiment did not fail due to the accuracy of the device or because of the data that was collected. The Michelson–Morley experiment failed because they used the wrong equation (or algorithm) to convert the data from raw measurements into Earth orbital velocities. We will examine the corrected algorithm in Chapter 7.

As you can see, a continued belief in relativity requires you to ignore Michelson and Morley's actual data and statistics.

Summary

By now, you should have an understanding of two of the most significant shifts in the history of physics: the first move was to establish classical mechanics, or first–generation physics. The second was the shift to modern physics, or second–generation physics, which encompasses relativity theory and quantum mechanics. You should also understand that a nonobvious theory requires four elements: assumptions, derivation, proof, and implications. Einstein's theory of relativity followed these steps and was believed to be free of any fatal mistakes. While it has withstood an onslaught of historical challenges, relativity theory suffers from a critical mistake. A key component, its spherical wave proof, failed, but did so in such a way that people believed it had actually passed. Said simply, Einstein thought he had a spherical wave when in actuality, he didn't. This is a Type I error, one that has proven to be extremely elusive. In addition to revealing a mistake in Einstein's theoretical foundation, we have mentioned that Modern Mechanics' equations provide better results than the relativity equations for the Michelson–Morley experiment.

A failure of the spherical wave proof along with new equations that provide better results, while indicative of a problem in Einstein's work, does not identify what that problem is. These findings raise several interesting and important questions:

1. *How can Einstein's work be wrong while still producing really good results in many classes of experiments?*

2. *If there is a mistake in Einstein's work, why hasn't it been previously discovered?*

3. *Why must Einstein's spherical wave proof work on all points of a sphere and not just specific, individual points?*

4. *Why do Einstein's equations perform worse than Modern Mechanics' equations?*

5. *Can Einstein's equations be fixed, and if not, what are the implication for physics and science?*

The answer to these questions will be revealed in the pages that follow. Before we continue examining Einstein's work, we must first develop the Modern Mechanics model. Chapters 2 through 4 develop the conceptual and mathematical framework for Modern Mechanics. We then use Modern Mechanics, along with an explanation of functions developed in Chapter 5, as a springboard to understand Einstein's work. Chapter 6 examines Einstein's relativity derivation, reviewing its key concepts – which make it unique – and critical mistakes, which make it wrong. Built upon the material developed in chapters 2 through 5, Chapter 6 answers many of the questions we have just raised. We then use Chapter 7 to review several experiments to examine why Einstein's equations perform better than their classical mechanics counterparts and why Einstein's equations do not perform as well as the Modern Mechanics' equations. Chapter 8 examines the implications of Modern Mechanics and identifies interesting research areas.

A key implication of these findings is that relativity theory is incorrect, cannot be corrected, and cannot continue to exist. The removal of relativity theory from modern physics would create a crisis if there were no alternative that would produce equal or better results. Fortunately, Modern Mechanics is a mathematically and conceptually sound alternative. Modern Mechanics begins with a new set of assumptions and rules, distinguishing it from first and second–generation models. The result is a different theory and set of equations that yield equal or better results than relativity theory. Modern Mechanics has an ambitious goal: to be a unified model that explains things in an intuitive way.

Chapter 2 Motion Involving Two Systems

Modern Mechanics has an ambitious goal: to serve as a unified model that explains motion. Similar to historical models, it describes and explains the behavior of moving systems. It builds upon the best of classical mechanics to explain the behaviors of *length* and *time*. Like relativity theory, it also explains the behaviors of *frequency* and *wavelength* in moving systems, but without the mistakes, constraints, and non–intuitive terminology that accompany Einstein's theory. It does all this with equal or better accuracy. At a minimum, it will eliminate the need for both classical mechanics and relativity theory. Ideally, it establishes a platform that will also unify the areas of physics currently covered by quantum mechanics.

Modern Mechanics is based on geometric transformations, a mathematical framework that is well defined and easy to understand. Because geometric transformations are very well understood, both mathematically and conceptually, some readers may feel the urge to skip ahead. Resist the temptation. The next two chapters will introduce several terms and concepts that will be used throughout the remainder of the book.

Modern Mechanics is friendly for people who are new to physics. While it is built from the ground up and introduces some new terminology, the essential concepts will be familiar to most readers. It combines ideas in modern thinking – in physics, mathematics, computer science, and engineering – to arrive at a model that can explain many things that we're familiar with today. Also, it is extensible, enabling it to explain things in the future that today are unknown.

Many readers will find geometric transformations familiar and intuitive, because they are based on mathematical principles that are taught in elementary school. However, many people do not realize they are using geometric transformations to answer "everyday questions." The Disneyland question from Chapter 1 is an example of a geometric transformation. Examples of questions that can be answered with geometric transformations are: How long have we been driving? How far have we gone? Where are we now? What direction are we facing? Are we going the right way?

We need a set of terms and concepts upon which we will build Modern Mechanics. The key concepts are surprisingly simple. Imagine for a moment each of the following: A woman walking down the street; a boy making a right–hand turn on a bicycle; a man stretching a rubber band in an open room, and a girl turning the page of a magazine on a table. When you were asked to imagine the different motions, there was always an **object** involved: the woman, the bicycle, the rubber band, the street, the room, the table, or the magazine. There was also an action or motion involved. Each of these motions is described with geometric transformations.

Physics would be a very challenging discipline if we needed a unique model or theory for every individual object we encountered. Fortunately, we can create generalized models that apply equally to a wide range of objects, or *things*. In Modern Mechanics these objects are all called **systems**. The woman, a bicycle, the rubber band, the street, the room, the table, and the magazine are all examples of systems.

A system, sometimes called a coordinate system, is a foundational concept within Modern Mechanics. Conceptually, **systems** are things that have one, two, or three dimensions. Systems are measured in terms of length or wavelength. Generally speaking throughout this book, a one–dimensional system is called a **line,** a two–dimensional system is called a **plane**, and a three–dimensional system is called a **space**. Example systems are illustrated in Figure 2–1.

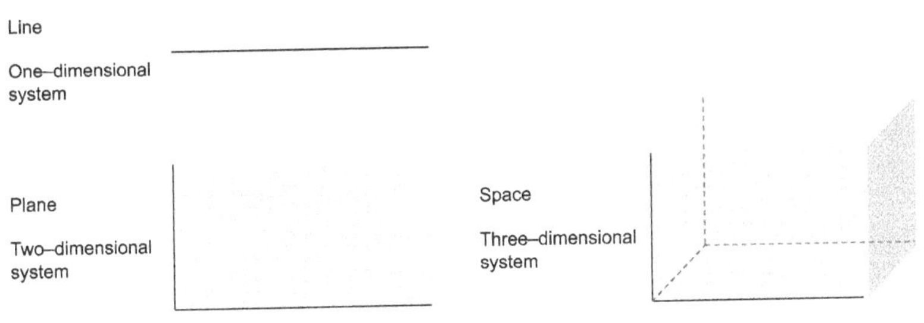

Figure 2–1 Illustrations of one, two, and three–dimensional systems. A one–dimensional object is shown as a line, a two–dimensional object is shown as a plane, and a three–dimensional object is shown as a space.

Systems are not very exciting when presented as images on a page, because they don't move. What makes them interesting is when we combine them with one another and then put them in motion. To accomplish this, Modern Mechanics uses four rules that define systems and their interactions:

1. A system is a one, two, or three–dimensional spatial coordinate system that is defined by its boundaries, its internal propagation medium(s), or both.

2. A system must be placed entirely on or in another system.

3. A system uses geometric transformations to move *with respect to* the system on or in which it was placed, and is viewed *from the perspective of* any other system.

4. A system observed by its boundaries can be thought of as a particle, while a system observed as a propagating medium(s) can be thought of as a wave.

We are able to develop a complete model that explains motion in terms of velocity, length, and time using just these four rules.

Rule 1*: A system is a one, two, or three–dimensional spatial coordinate system that is defined by its boundaries, its internal propagation medium(s), or both.*

Rule 1 defines a system. Systems can be definite, or finite, where their dimensions are known. They can also be indefinite, or infinite, where their size is unknown. While systems with defined boundaries are used throughout this book, it is important to note that in some systems the characteristics of a system's medium(s) are more important than its boundaries. As a result of this distinction, there are several nuances that must be immediately addressed. First, in Modern Mechanics, a point is a special kind of system. It is an abstract type that has no dimensions. While it represents a position, it has no physical size. This is consistent with how points are treated in mathematics. This is an important nuance that must be addressed, since an object, such as a particle, might sometimes be graphed as a point. However, while a particle has dimensions and can exist in the physical world, a point cannot. Second, a system can contain or consist of multiple mediums, with each medium possibly having its own unique characteristics (eg, how fast a wave travels through it). Multiple mediums can exist simultaneously. Also, a physical boundary may not always be sufficient to separate the medium of one system from that of another.

For example, a swimming pool can contain water, which is one type of medium. Simultaneously, it can contain an electromagnetic aether, which is another type of medium. Thus, when talking about a pool, we can talk about a particle traveling through the pool, a sound wave traveling through the pool, or a beam of light traveling through the pool. This association of a medium (or mediums) with a system will become important when we consider the behavior of waves and particles. This definition of a system helps explain the existence of the fourth rule, given that emphasis can be placed on a system's boundaries, its medium(s), or both.

Rule 2*: A system must be placed entirely on or in another system.*

Systems that other systems are placed on or in are called **outer systems**. The first system we place is the outermost system. As you will learn in Chapter 3, where we discuss nested relationships, there can be more than one outer system, but there will only be one outermost system. For simplicity, we use "outer system" to mean the outermost system without complicating matters by referring to more than one outer system. By itself, an outer system isn't very exciting, because when it is the only system, there isn't anything it can move with respect to. Additionally, as the only system, there isn't anything on or in it that can be moved. However, an outer system is critical to motion because it gives us an initial *reference system* that we can use to calibrate and measure distance and time. An outer system is illustrated in Figure 2–2. Reference units should be considered part of their respective systems, not system themselves. Without this reference system, there is no concept of absolute time or distance, both of which are required for motion to occur. As an example, consider a freeway full of cars. The freeway would be the outer system upon which all of the cars are placed.

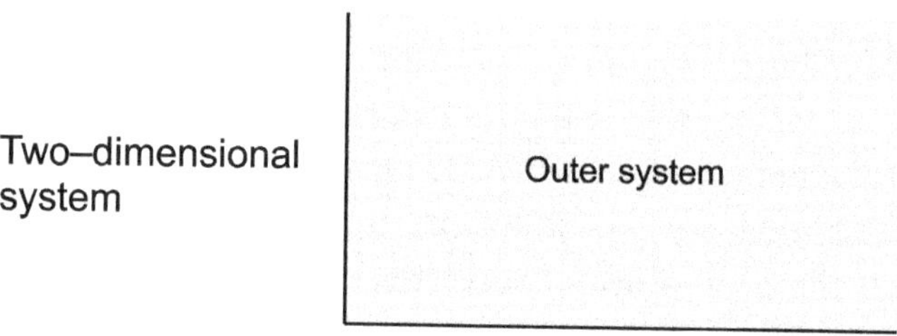

Figure 2–2 A two–dimensional outer system. Outer systems serve as a reference system for other systems that are placed on or in them.

Rule 2 states that systems must be placed on or in other systems, with a notable exception being the placement of the first system. Beginning with the second system, all systems must now be placed on a previously placed system according to the rule. The second rule is what enables systems to relate to and interact with one another. The interaction of two or more systems is called a **model** or a **relationship**. As illustrated in Figure 2–3, systems that are placed on or in other systems are called **inner systems**. Inner systems are interesting because they can move *with respect to* the system on which they were placed. As a reference for measuring distance, we arbitrarily set the bottom–left corner of each system as its origin throughout this book. Continuing with the freeway example, the cars placed on the freeway are all individual inner systems.

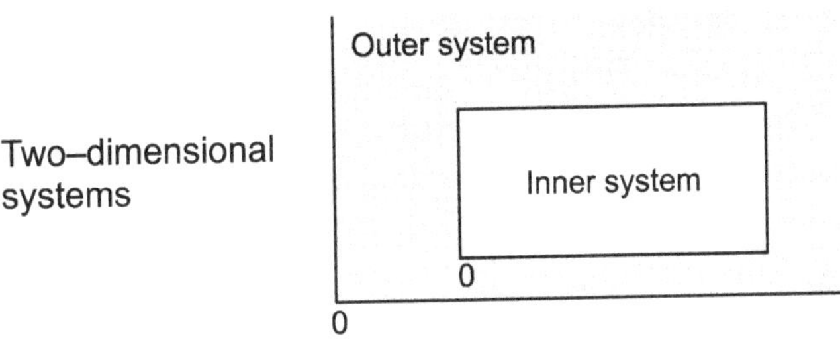

Figure 2–3 Placement of an inner system on or in an outer system. The inner system moves *with respect to* the outer system.

An inner system's motion and behavior is made "*with respect to*" the system on or in which it was placed. This phrase is important and must be distinguished from the measurement of that inner system's motion or size from the point of view of any other system, where it would be described "*from the perspective of*" that other system. This important nuance in terminology enables us to identify which system is the reference system and which systems might have a distorted view of that inner system's motion. Generally, when a system's position is measured *with respect to* the outer system on which it was placed, we use *absolute* distance measurements. When a position is measured from *the perspective of* another inner system, we use *relative* distance measurements. As an example, when two cars are driving parallel to one another at 60 miles per hour *with respect to* the freeway, then *from the perspective of* each vehicle, the other vehicle appears stationary.

Rule 3: *A system uses geometric transformations to move with respect to the system on or in which it was placed, and is viewed from the perspective of any other system.*

Rule 3 allows systems to move. We begin by describing the motion of two systems: an outer system and an inner system. The movement of an inner system occurs *with respect to* an outer system. This means that we use the outer system as our reference system for determining placement, position, and orientation of the inner system. The motion of the inner system is governed by rules of geometric transformations, or simply transformations. To help illustrate the various ways in which transformations occur, imagine for a moment a Slinky, which is a spiral–shaped, spring–like toy. Further, imagine it lying on its side on the table. How many different ways can you move the Slinky on a table? You can slide it across the table. You can spin, or rotate it. You can flip it. And, you can stretch it. Using the language of geometric transformations, these motions are called translation, rotation, reflection, and scaling, terms that form the foundation of Modern Mechanics. While Modern Mechanics recognizes the importance of all geometric transformations, translation and scaling play the most prominent roles.

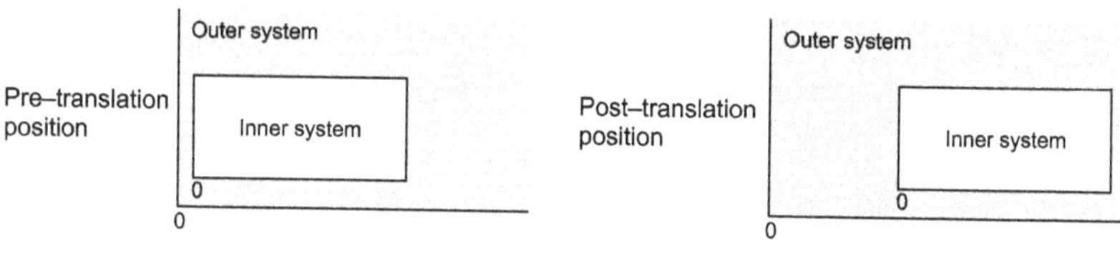

Figure 2–4 Translation. The inner system appears to slide along the outer system. The inner system begins in one position and then moves into a new position in the outer system, without changing size or orientation.

First, an inner system can slide, which is called *translation*. As illustrated in Figure 2–4, translation is defined by a change in the position of the inner system with respect to the outer system. When the inner system moves, its position with respect to the outer system changes, but it retains its original orientation (the way it is facing) and its original size. Translations are performed by adding a specific value to the initial position of the inner system, with respect to the outer system, to place it in a new position. For example, if the original position of the lower–right corner of the inner system was at x and it was moved d units to the right, the new position of the inner system would now be x'. In a general sense, this movement is written as:

$$x' = x + d$$

Often in physics, we will want d, or the distance that the inner system moves, to vary with time. When written to reflect the passage of time, the equation becomes the Newtonian transformation, which is written as:

$$x' = x + vt$$

where v is the velocity of the inner system and t is the time it has been in motion. In Modern Mechanics, d is considered *static* because it does not vary with time, whereas vt is considered *dynamic* because it varies with time. The Newtonian transformation is one of the most important transformations in both classical mechanics and Modern Mechanics. In fact, we used a variant of this equation to answer the question in Chapter 1 regarding how long it would take to drive to Disneyland.

Translations can occur in any direction, although in the examples that we will develop throughout this book, the inner system will "slide," or move, in a straight line to the left or to the right. An example of a translation is a woman (inner system) *walking in a straight line* (translation) down a street (outer system).

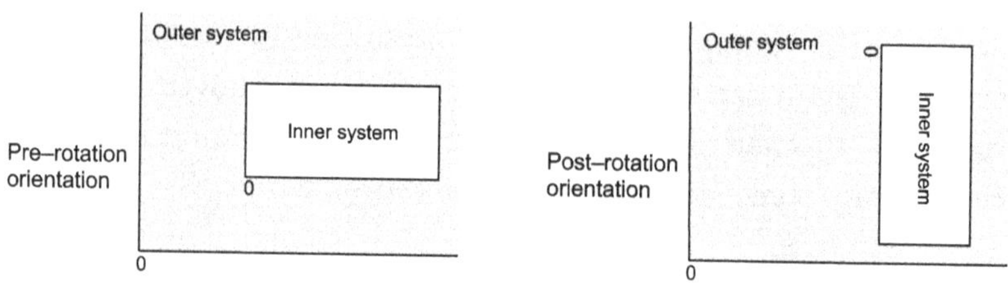

Figure 2–5 Rotation. The inner system turns with respect to the outer system. With rotation, the inner system changes its orientation with respect to the outer system. It does not change in size. However, the inner system may end up in a new location, dependent on the center point and axes of rotation.

Second, an inner system can turn, which is called *rotation*. Illustrated in Figure 2–5, rotation is a change in orientation, or a change in the angle or direction, of the inner system with respect to the outer system. The inner system faces a different direction, but the original size and shape remain the same. With rotation, the final position of the inner system *may* change and is dependent upon the center point that is used when the rotation is performed. Rotations can occur along any axis and in any direction. Rotations can be referred to as clockwise and counterclockwise. The equation for each transformed point in three dimensions is:

$$x' = x\cos\theta - y\sin\theta$$
$$y' = x\sin\theta + y\cos\theta$$

An example of rotation is a boy riding a bicycle (inner system) making a *right turn* (rotation) on a street (outer system).

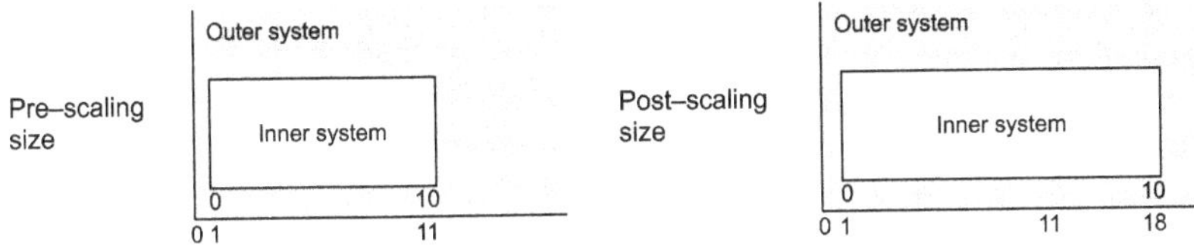

Figure 2–6 Scaling. The inner system appears to stretch with respect to the outer system.

Third, an inner system can stretch, which is called *scaling.* As illustrated in Figure 2–6, scaling is the change in size of the inner system with respect to the outer system. The inner system is "stretched," but retains its original orientation and at least one of its starting points. The inner system can be stretched along one or more axes. Measured with respect to the outer system, the inner system has stretched and has a new size (length). *From the perspective of* the inner system, nothing has changed. This isn't as intuitive as it sounds because, looking at Figure 2–6 with respect to the outer system, you can see the size has physically changed. This distinction occurs because of a difference between absolute and relative measurements.

Consider an example of a balloon (inner system) that is 4*cm* wide and is sitting on top of a ruler (outer system). While it is deflated, draw three evenly spaced lines around the balloon, effectively dividing it into four sections of equal length. The distance between the lines is a length of 1 unit according to both the inner system and outer system. However, in the outer system, the units are called centimeters. Now blow air into the balloon. *With respect to* the outer system, the inner system's (ie, balloon's) size has changed (in terms of centimeters), while *from the perspective of* the inner system there are still four equal quarters. In this example, the outer system forms an absolute measurement frame where the length between the marks has increased; while the inner system forms a relative measurement frame, where the distance between the lines is still 1 unit. An example of scaling is a man stretching (scaling) a rubber band (inner system) in an open room (outer system). Mathematically, scaling is generally performed using multiplication or division.

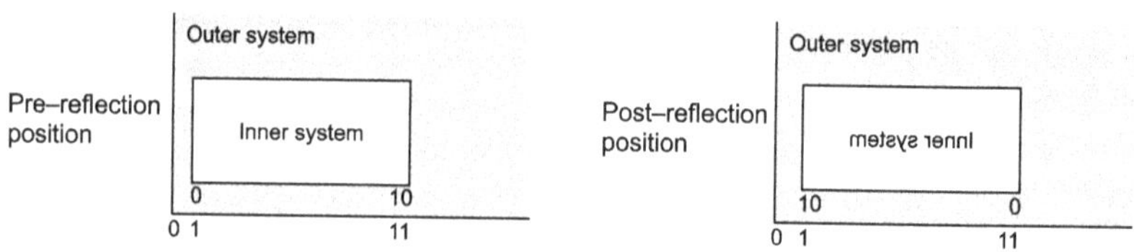

Figure 2–7 Reflection. The inner system is flipped with respect to the outer system.

Fourth, as shown in Figure 2–7, an inner system can flip, which we call *reflection*. From the perspective of the outer system, the inner system has flipped along one axis. One way to think of this is similar to the turning of a page in a book. An example of reflection would be a girl flipping (reflection) a magazine page (inner system) on a table (outer system).

Transformations can be performed one at a time, or in combination with one another. For example, anyone who has ridden an amusement park ride has experienced the combination of translation and rotation on many roller coasters. The transformations rotation, translation, and reflection retain the inner system's original size and shape, and are called *congruent*. Because scaling can change the size and/or shape of the inner system with respect to the outer system, it is called *similar*. First and second–generation physics emphasize congruent transformations. Modern Mechanics recognizes the importance of both congruent and similar transformations.

This chapter introduced geometric transformations to a sufficient level that allows us to develop Modern Mechanics. It did not present a detailed mathematical treatment of geometric transformations. The concepts and mathematics surrounding the geometric transformations of translation, rotation, scaling, and reflection are well developed and understood. For additional material and a detailed mathematical treatment of transformations and their associated equations, a recommended resource is the book *Computer Graphics: Principles and Practice*, which in 2013 was in its third edition. Additionally, this material is often presented in high–school and college–level geometry books.

Rule 4: *A system observed by its boundaries can be thought of as a particle, while a system observed as a propagating medium(s) can be thought of as a wave.*

The fourth rule allows us to recognize and distinguish between particles and waves. Interestingly, a system can begin as a particle and later participate as part of a wave. For example, a water droplet can be thought of as a particle when it is falling from the sky, but then can be part of a wave when it takes its place in a lake.

The behavior of particles is explained using one of the most important equations in physics, which defines the relationship between velocity, v, time, t, and length, l:

$$v = \frac{l}{t} \qquad \text{Eq. 2.1}$$

In Modern Mechanics, we refer to Equation 2.1 as the length–based motion equation, or simply the **length equation**. In a general sense, the length equation is the foundational equation to explain the motion of particles. It is also the foundation of classical mechanics and describes motion quite well. As will be discussed shortly, length l and time t in Equation 2.1 are both discrete types, while velocity v is a compound type.

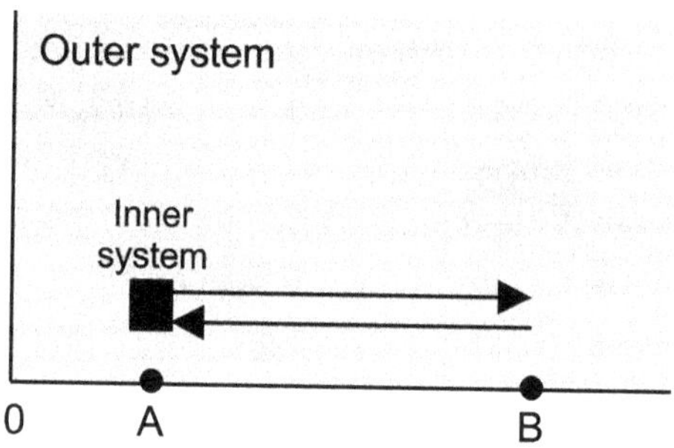

Figure 2–8 The inner system, represented by the black square, moves from point A on the outer system to point B, then back to point A. The repeated back–and–forth motion is called oscillation.

Compound types differ from discrete types because the former can be used to explain cyclical, or back–and–forth, motion. In a two–system model, oscillation requires us to introduce two points, A and B, which are placed at fixed positions. In this example, the points are placed on the outer system. (Technically, points are systems, however to enhance readability, we will refer to them as points and will not treat them as distinct systems). The inner system is placed into back–and–forth motion between these two points, as illustrated in Figure 2–8. This cyclical motion is called **oscillation**.

Conceptually, oscillation consists of two motions: one from point *A* to point *B*, and a second from point *B* to point *A*. Each one of these motions is called **a segment** and the combination of two segments is called a **cycle**. When the oscillating system is moving away from the origin (eg, from point A toward point B), this is called the **forward segment**. When the oscillating system is moving toward the origin (eg, from point B to point A), it is called the **reflected segment**.

A key characteristic of wavelength that distinguishes it from length is that wavelength is bidirectional. It starts at point A and moves in one direction to arrive at point B, where it reverses direction and returns to point A. To prevent erroneous or nonsensical answers associated with bidirectional movement, care must be taken when performing mathematical operations on wavelength. Depending on the problem to be solved, you may need to average two segments rather than take their sum or difference.

When dealing with oscillation, or cyclical motion, we can ask interesting questions, such as: How long does it take to make a round–trip journey? What is the total distance of a round–trip journey? Because a round–trip journey occurs in two segments, how long is each segment? How long does it take to traverse each segment? Because a round–trip journey occurs in two segments, what is the average length of the two segments? Because a round–trip journey occurs in two segments, what is the average segment traversal time?

Motion involving oscillation is expressed in terms of velocity, v, frequency, f, and wavelength, λ, where the relationship is defined as:

$$v = f * \lambda \qquad \text{Eq. 2.2}$$

This is called the wavelength–based motion equation, or simply the **wavelength equation**. In a general sense, the wavelength equation is the foundational equation that explains the motion of waves.

Mathematically, Modern Mechanics relies on the proper treatment of types, which fall into two categories: discrete and compound. Discrete types do not change in value due to a change in another value. For example, if you measure 10 meters on the ground and place cones at the start and end of this length, the length between the cones does not change with the passage of time. A **discrete type** is simply a value that is static and does not vary with respect to something else.

A **compound type** consists of two values, where part of the value varies with respect to something else. Compound types are generally written in terms of two other types separated by the word "per." For example, kilometers per hour is a compound type consisting of two discrete types: kilometers and hours. We refer to the type preceding the word per as the **type of the numerator** and the word following the word per as the **type of the denominator**.

Differing from discrete types; compound types consist of a type of the numerator and a type of the denominator. As an example, if meters is the type of the numerator and seconds is the type of the denominator, then the compound type is meters per second. Because compound types are rates, the value of its numerator cannot be found unless we first know the value of the denominator. For example, if a vehicle is moving at 40km/h, we cannot determine how many kilometers it has traveled without first knowing how many hours it has been traveling. Notice that "40 kilometers" and "40 kilometers *per* hour" represent different types and are not interchangeable.

There are three fundamental *discrete types* in Modern Mechanics: **distance**, **cycle**, and **time**. Distance is the measurement of constant intervals of length between two points. A cycle is a count of repetition. It is the measurement of constant intervals of repetitions or cycles. For example, a spring might consist of 10 winds of a coil (eg, 10 cycles). Time is a measure of motion. Specifically, time is the measurement of constant intervals of motion. Time, like distance and cycles, is a discrete type: It is static and independent of the passage of time. This may sound counterintuitive, and one might ask: How can time be independent of time? Consider that the time interval on the face of your wristwatch between the 12 and the 1 represents five seconds and will always represent five seconds, regardless of how many seconds have passed.

There are three fundamental *compound types* in Modern Mechanics: **velocity**, **wavelength**, and **frequency**. As discussed previously, these are compound types that are built using the three basic discrete types of distance, cycles, and time. Compound types can be developed using discrete types, compound types, or their combinations.

Each type can also be inverted by simply dividing it by one, resulting in inversed distance, inversed cycles, inversed time, inversed velocity, inversed wavelength, or inversed frequency.

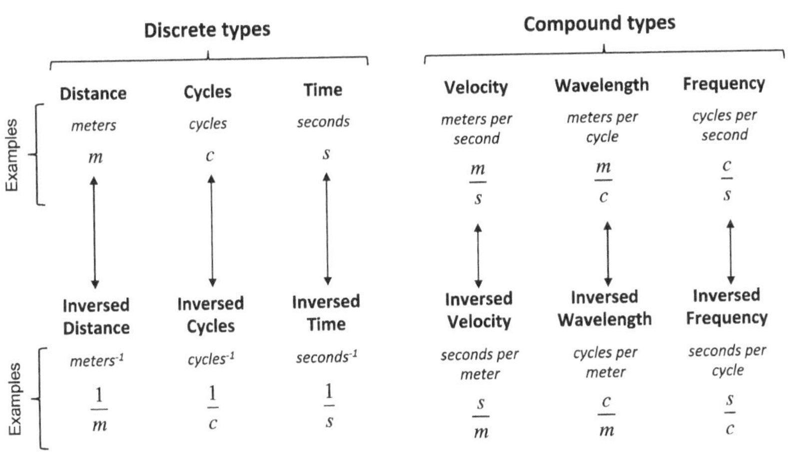

Figure 2–9 Modern Mechanics' type and inverse type relationships for discrete and compound types.

It is important to recognize that discrete and compound types are different. A common mistake is to mistreat wavelength, which is a compound type and is defined as *length <u>per</u> cycle*, as if it were simply a *length*, which is a discrete type. This is no different than reading a speedometer as kilometers instead of as kilometers per hour. As you will read in later chapters, this distinction between discrete and compound types is an extremely subtle and important differentiator between Modern Mechanics and its counterparts.

Summary

This chapter introduced the key concepts and characteristics of Modern Mechanics. Grounded in geometric transformations, two–system models in Modern Mechanics describe motion in terms of velocity, length, and time. A two–system model is similar to its first–generation, classical mechanics cousin. Both explain aspects of motion using geometric transformations. The distinguishing characteristics of Modern Mechanics include the use of length and wavelength equations, the correct treatment of types, and the recognized use of averages.

Modern Mechanics is built upon four rules that define systems and their relationship with other systems.

1. A system is a one, two, or three–dimensional spatial coordinate system that is defined by its boundaries, its internal propagation medium(s), or both. Systems can be finite or infinite.

2. A system must be placed entirely on or in another system.

3. A system uses geometric transformations to move *with respect to* the system on or in which it was placed, and is viewed *from the perspective of* any other system.

4. A system observed by its boundaries can be thought of as a particle, while a system observed as a propagating medium(s) can be thought of as a wave.

Classical mechanics was useful to explain motion in terms of time, velocity, and length. However, it was unable to satisfactorily explain observations and experiments – such as the Michelson–Morley experiment – involving velocity, wavelength and frequency. The inability of classical mechanics to explain oscillation is not due to a limitation of geometric transformations, but is simply a characteristic of a two–system model. Modern Mechanics recognizes that frequency, velocity, and wavelength cannot be addressed using a two–system model. Describing oscillating motion is addressed in Modern Mechanics with the introduction of a fifth rule and a third system. As a unified model, Modern Mechanics builds upon the principles of motion defined by geometric transformations.

Chapter 3 Motion Involving Three Systems

Modern Mechanics is a unified model that explains motion. Like classical mechanics, Modern Mechanics is highly effective at explaining motion and behavior in terms of velocity, time, and length. Until the advent of experiments involving electromagnetic force and optics, the classical–mechanics equations were the most accurate available. However, light and radio waves did not appear to always behave according to the equations for time, length and velocity. In fact, the classical–mechanics equations performed so poorly in explaining cyclical behaviors that they could not be used in a consistently valuable or useful manner. This was a significant limitation that classical mechanics was unable to overcome.

As introduced in Chapter 2, a one–system model enables us to describe a system in terms of its boundaries, its space, or both. However, a one–system model is rather uninteresting, since no motion occurs without a second system. A two–system model is interesting because it allows us to explain motion using the spatial relationship between an outer system and an inner system.

In a two–system model, the inner system moves with respect to the outer system. Two–system models are extremely useful in explaining motion in terms of *velocity*, *length*, and *time*. However, two–system models are unable to explain the behavior of *velocity*, *wavelength*, and *frequency* involving moving systems. A two–system geometric transformation–based model cannot explain the complex and interesting oscillations that occur with moving systems.

The wavelength equation is fundamentally about cyclical behavior, and no two–system geometric transformation–based model is able to explain this behavior in moving systems. The inability to explain this behavior created a crisis in 19th century physics. To overcome the problem, Einstein introduced an alternative two–system theory called relativity. Instead of relying on geometric transformations, relativity theory relies on two principles: 1) *the principle of the constancy of the velocity of light* and 2) *the principle of relativity*. As we have shown in Chapter 1, the proof establishing the compatibility between these two principles failed. This creates a new crisis, since we need a model to explain the experiments that were previously explained by relativity theory.

Modern Mechanics is unique because it uses geometric transformations to explain the behavior of wavelengths and frequencies in moving systems. Modern Mechanics answers questions of *velocity*, *frequency*, and *wavelength* by introducing a third system. Many readers will find that they are inherently familiar with the key concepts of Modern Mechanics because they are well–grounded in mathematics and physics.

This chapter builds upon the foundational concepts of Modern Mechanics: systems, geometric transformations, and *the four rules* that define the relationship between systems. To this foundation, Modern Mechanics adds a new rule; one specific to oscillation:

***Rule 5**: Cyclical behavior is explained using three–system models, where the third system, called the oscillating system, oscillates between two points on the inner system.*

On the surface, Rule 5 sounds extremely simple. However, it introduces many complexities that must be explained to fully understand and appreciate its power. Three–system models start as two–system models, consisting of an outer system and an inner system. We use Rule 2 – *A system must be placed entirely on or in another system* – to add a third system to a two–system model. Technically, we are introducing a second inner system. However, to avoid name confusion we will refer to this new system as the **oscillating system**.

Notice that the placement of the oscillating system requires a decision. Rule 2 enables us to place the oscillating system on any previously placed system. Therefore, it can be placed on the existing outer system or it can be placed on the previously placed inner system. Regardless of the system upon which the oscillating system is placed, it oscillates between two points on the inner system. The placement of the oscillating system on either the outer system or on the existing inner system defines how the model and equations behave.

Non–Nested System Relationships

When the oscillating system is *placed on or in an outer system,* it creates a **non–nested system relationship**, or simply a non–nested relationship. A non–nested relationship is illustrated in Figure 3–1, where the newly placed oscillating system is shown as a black square.

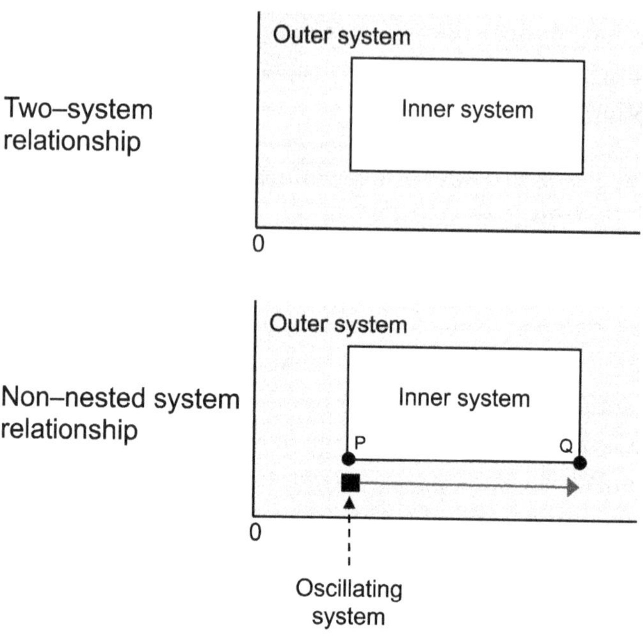

Figure 3–1 A non–nested system relationship consisting of two non–nested inner systems. Both the inner system and the newly placed oscillating system (shown as a black square) are placed on and move with respect to the outer system.

In Figure 3–1, first consider the **forward segment** of the oscillating system's motion as it moves from point P to point Q, which are both points *on the inner system*. The arrow represents the direction of motion of the oscillating system on segment $\overrightarrow{PQ}$. In this example of a non–nested relationship, the inner system is motionless, while the newly placed oscillating system is in motion. Later in this chapter, we will develop the behavior when the inner system is in motion. In a non–nested relationship, the oscillating system moves *with respect to* the outer system (since this is the system on which it was placed). The oscillating system can also be observed *from the perspective of* the inner system. The distinction between *"with respect to"* and *"from the perspective of"* is an important characteristic of Modern Mechanics.

As a real–world example of a non–nested relationship, consider a moving sidewalk (inner system) that you might encounter while walking along a hallway (outer system) at an airport. You (oscillating system) choose to walk along the hallway (outer system) instead of taking the moving sidewalk.

Nested System Relationships

When the oscillating system is *placed on or in an inner system* it creates a **nested system relationship**, or simply a nested relationship. A nested relationship is illustrated in Figure 3–2, where the newly placed oscillating system is shown as a black square.

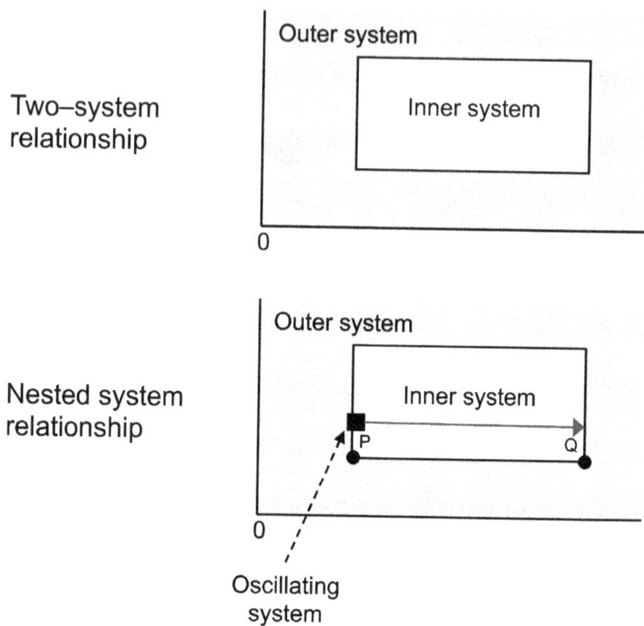

Figure 3–2 A nested system relationship consisting of two nested inner systems. The newly placed oscillating system (shown as a black square) is placed on the inner system, which was previously placed on the outer system. The oscillating system moves with respect to the inner system and the inner system moves with respect to the outer system.

In Figure 3–2, consider the forward segment of the oscillating system's motion as it moves from point P to point Q, which are both points *on the inner system*. The arrow represents the direction of motion of the oscillating system on segment $\overrightarrow{PQ}$. In this example of a nested relationship, the inner system is motionless, while the newly placed oscillating system is in motion. We will later develop the case when the inner system is also in motion. The oscillating system moves *with respect to* the inner system (since this is the system on which it was placed). The oscillating system can also be observed *from the perspective of* the outer system. In a nested relationship, the inner system takes on a dual role: It is an inner system with respect to the outermost system, and it also serves as the outer system for the newly placed oscillating system.

As a real–world example, once again consider the moving sidewalk (inner system) that one might encounter while walking along a hallway (outer system) at an airport. You (oscillating system) choose to take the moving sidewalk where, even if you are standing still *with respect to* the moving sidewalk, you appear to be moving *from the perspective of* someone in the hallway. Notice that the moving sidewalk takes on the role of *your* outer system, since that is what you are moving *with respect to* and it is an inner system *with respect to* the hallway, creating the nested relationship.

The motion of oscillating systems is described using the same geometric transformations used to describe the motion of inner systems. However, because of the cyclical motion of an oscillating system, we must consider several unique characteristics. The most significant is bidirectional travel; meaning the oscillating system goes in one direction for the forward segment and then goes in the other direction for the reflected segment. As a result, we cannot always use addition or subtraction when describing the behavior of the oscillating system or any resulting waves it might produce. We will sometimes add segments and at other times we will use their average. One of the most important equations associated with oscillation and Modern Mechanics is the average.

In summary, when an oscillating system is placed on an outer system, it creates a non–nested relationship. When an oscillating system is placed on an inner system, it creates a nested relationship. *Regardless of whether the oscillating system is placed on the inner system or on the outer system, it oscillates between points P and Q, which are both on the inner system.*

Motion Involving Two Moving Systems

Explaining the behavior of particles and waves in terms of *velocity*, *frequency*, and *wavelength* requires that the inner system and the oscillating system both be in motion. The inner system is placed in translatory motion with respect to the outer system and moves in one direction only. Translatory motion simply means that it continues in a straight line at a constant velocity. In our examples, the inner system moves to the right, with respect to its outer system. The oscillating system oscillates, or moves back and forth, with respect to the system on which it was placed. These behaviors of the inner and oscillating systems are illustrated in Figure 3–3 for both nested and non–nested system relationships.

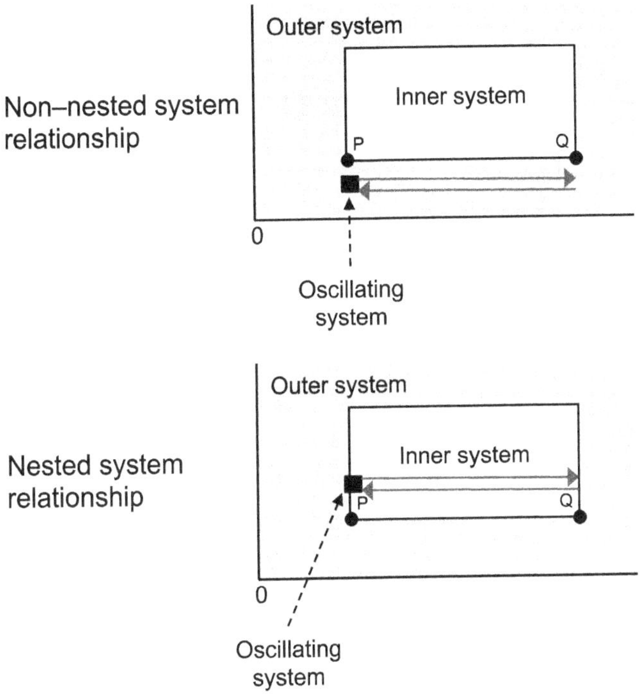

Figure 3–3 Nested and non–nested relationships. The oscillating system is shown with its forward and reflected segments. The oscillating system moves back and forth between points P and Q of the inner system. The inner system is not in motion. In a nested relationship, the oscillating system moves with respect to the inner system. In a non–nested relationship, the oscillating system moves with respect to the outer system.

The oscillating system moves along the forward segment $\overrightarrow{PQ}$. After reaching point P, it then moves along the reflected segment $\overleftarrow{PQ}$ and travels back to point P. This round–trip journey from point P to point Q, returning to point P, is called a **cycle** or **oscillation**. The distance of and time required for the oscillating system to travel the forward segment, $\overrightarrow{PQ}$, are called the **forward intercept length** and the **forward intercept time**, respectively. The distance of and time required for the oscillating system to travel the reflected segment, $\overleftarrow{PQ}$, are called the **reflected intercept length** and the **reflected intercept time**, respectively. When the context is obvious, the words "forward" and "reflected" are dropped and these terms may simply be referred to as the **intercept lengths** and **intercept times**. We are also able to determine the **average intercept length** and **average intercept time** by using the forward and reflected intercepts.

Cyclical motion can describe the motion of particles and waves, but also plays an important role in describing the relationship between frequency, wavelength, and velocity in moving systems. The simplest case occurs when the oscillating system behaves like a particle as part of a nested relationship, as illustrated in Figure 3–4.

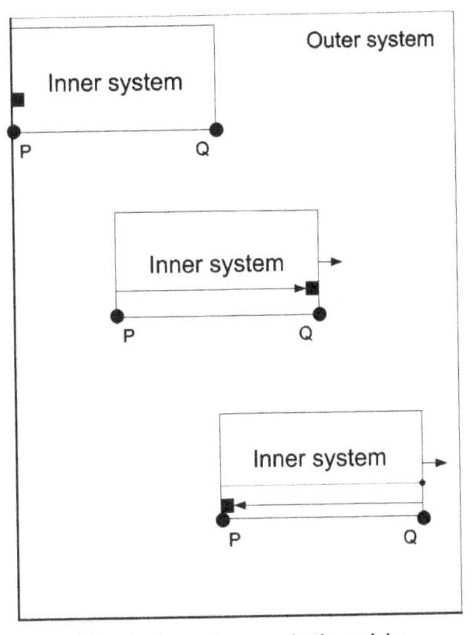

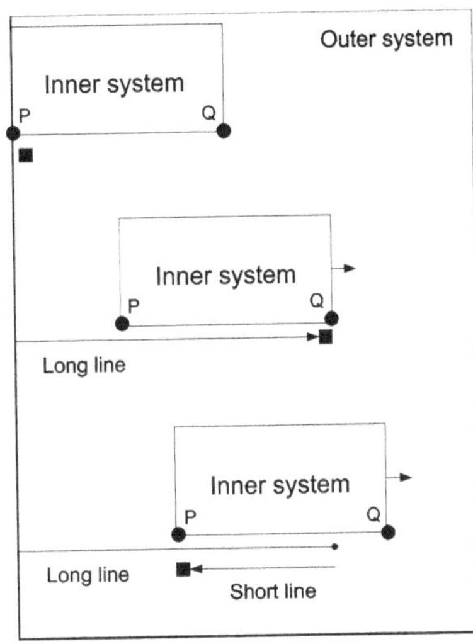

Figure 3–4 Both the inner system and the oscillating system are in motion. The inner system and the oscillating system can have different velocities and directions, with each moving at its own velocity, w for the oscillating system and v for the inner system. In a nested relationship, the oscillating system can complete both the forward and reflected segments, regardless of the inner system's velocity. In a non–nested system relationship, if the velocity of the inner system meets or exceeds that of the oscillating system, the oscillating system will never complete the forward segment.

The left side of Figure 3–4 illustrates a nested relationship, in which both the inner system and oscillating system are in motion. *In a nested relationship, the oscillating system moves with respect to the inner system.* Point P is the origin of the inner system and point Q is at the lower–right corner. The origins of the oscillating and inner systems begin coincident with each other. The oscillating system moves along the path of the forward segment, from the origin (point P) to point Q at the other end of the inner system. Upon reaching point Q, the oscillating system then reverses direction and moves along the path of the reflected segment, traveling back to point P at the origin *of the inner system.*

In a nested relationship, the forward and reflected intercept lengths are the same. This type of back–and–forth motion, where both intercepts have the same length, is called **symmetric oscillation**. *From the perspective of* the outer system, the apparent velocity of the oscillating system varies depending on its direction of travel: The oscillating system *appears* to travel faster when the inner system and the oscillating system are moving in the same direction, and slower when the inner system and oscillating system are moving in opposite directions. Remember, the actual motion of the oscillating system is determined *with respect to* the system on which it was placed. Since the oscillating system was placed on the inner system, the velocity of the oscillating system and the distance traveled on both segments is the same.

Notice that the oscillating system is able to complete a cycle, or oscillation, as long as it has a positive velocity. Also notice that there is no upper limit to the velocity of either the inner or oscillating system in a nested relationship. As a real–world example, visualize a person (oscillating system) running in the aisle of a bus (inner system) between its rear window (point P) and front windshield (point Q), while the bus is moving along a road (outer system). The person can travel between points P and Q regardless of the velocity of the bus.

The right side of Figure 3–4 illustrates a non–nested relationship, in which both the inner system and oscillating system are in motion. *In a non–nested relationship, the oscillating system moves with respect to the outer system.* The oscillating system begins coincident with the origin (point P) of the inner system and travels along the path of the forward segment to point Q at the other side of the inner system. Once it reaches point Q, the oscillating system then reverses direction and travels along the path of the reflected segment, arriving at point P at the origin of the inner system. Because the inner system is also in motion, the oscillating system travels further and takes more time to complete the forward segment, $\overrightarrow{PQ}$, than it does to complete the reflected segment, $\overleftarrow{PQ}$. Because the oscillating system segments are not of the same length, this type of back–and–forth motion is called **asymmetric oscillation**. *From the perspective of the inner system*, the transit time and velocity of the oscillating system appear to vary depending on its direction of travel, while the distance traveled *appears* to remain the same.

In a non–nested relationship, *if oscillations are required,* then the velocity of the inner system cannot meet or exceed that of the oscillating system, otherwise the oscillating system will never be able to complete the forward segment, $\overrightarrow{PQ}$. The oscillating system is only able to complete a round–trip journey in a non–nested relationship when the velocity of the inner system is less than that of the oscillating system. As a real–world example, visualize a person (oscillating system) running on a road (outer system) next to a bus (inner system) between its rear window (point P) and front windshield (point Q), where the bus is also moving along the road (outer system). The person can only travel from the rear of the bus (point P) to the front of the bus (point Q) when the velocity of the bus is less than that of the person.

The *position* of the oscillating system – with respect to the system on which it was placed – when it intersects point Q of the inner system is called the **forward intercept position**, or simply the **intercept position**. We also define a **reflected intercept position**, which occurs when the oscillating system traveling along the path of the reflected segment arrives at point P. In Modern Mechanics, it is all–important to recognize the differences between intercept lengths, intercept times, and intercept positions.

There are several important observations we need to make about nested and non–nested relationships. In a nested relationship, when the oscillating system has completed exactly one–half of its full oscillation, it has reached the other end of the inner system. In other words, the average intercept length of the oscillating system is the same as the lengths of the forward and reflected segments. In a non–nested relationship, the forward segment is longer than the reflected segment. As a result, when the oscillating system has traveled the average intercept length, it has not yet reached point Q of the inner system, which means it had not yet reached its intercept point.

In Modern Mechanics, oscillating systems explain the behavior of waves in terms of wavelength. While technically a wave can be thought of as its own system, in Modern Mechanics we describe a wave in terms of *multiple* rays. This is a key difference between a particle and a wave, where a particle is viewed as the motion of a single ray. For simplicity, a wave is simply a circular (two–dimensional) or spherical (three–dimensional) surface that is emitted from a specific point and expands in all directions at a constant velocity, w. The wave consists of an infinite number of rays. To further aid in understanding wave motion, we will now refer to the oscillating system as the **oscillating wave source** to reflect its multi–ray nature.

Following a similar approach as with the behavior of particles, we begin by examining the wave motion in a nested relationship, in which both the inner system and oscillating system are in motion. In a nested relationship, *the oscillating wave source and the wave medium move with respect to the inner system*. The general case is illustrated in figures 3–5 through 3–7.

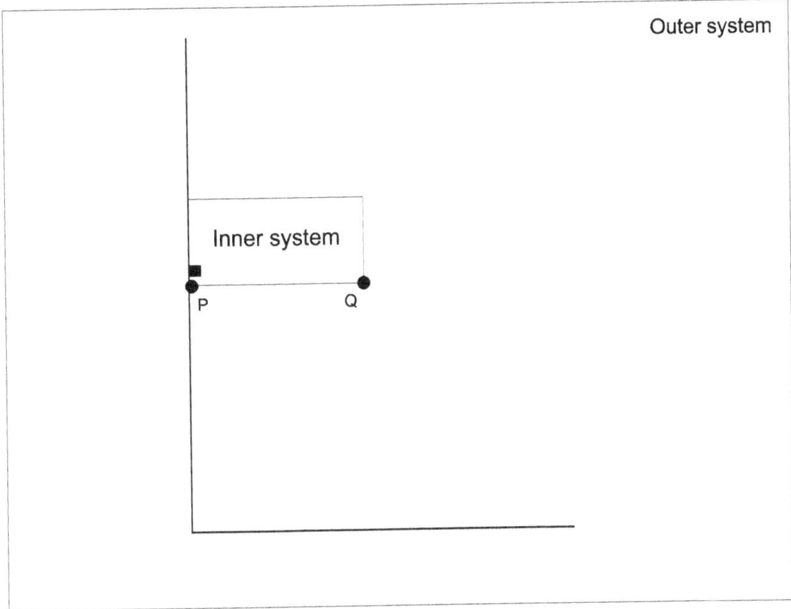

Nested system relationship
Symmetric oscillation

Figure 3–5 In a nested relationship at time $t = 0$ the emission of the spherical wave is coincident with the origin of the inner system. The size of the spherical wave (or bubble) gets larger with the passage of time. The oscillating system, represented by the black square, moves with respect to the inner system, which itself is in motion.

Points P and Q are on the inner system, with P at the origin and Q at the lower–right corner. The oscillating wave source begins coincident with the origin (point P) of the inner system and travels with respect to the inner system. The oscillating wave source begins at the origin and the rays move outward in all directions with respect to the inner system. It is best to visualize this behavior without regard to the outer system. The specific ray we will consider is the one that travels directly to point Q.

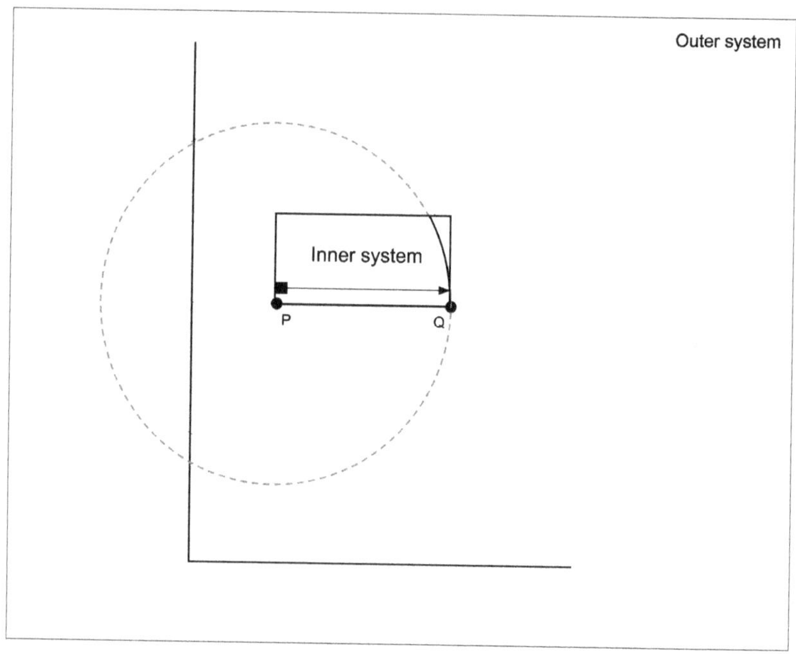

Nested system relationship
Symmetric oscillation

Figure 3–6 A nested system relationship after the oscillating system has reached point Q of the inner system. Because the oscillating system moves with respect to the inner system, the place in the inner system where the forward segment intercepts point Q is called the intercept point. In a nested relationship, this is simply point Q. Technically, the wave does not extend into the outer system. It is shown that way in the diagram so that you can see the circular (2D) or spherical (3D) waveform. Notice that the center of the circle is point P of the inner system.

As illustrated in Figure 3–6, the wave is depicted as a circle *centered at the origin of the inner system*. The oscillating wave source begins at the origin of the inner system and travels outward in all directions at the same rate. The ray we consider is the one that travels along the path of the forward segment to point Q at the other end of the inner system. Since this is a nested relationship, the oscillating system is able to travel from *P* to *Q* regardless of the velocity of the inner system. In the diagram, the circle is used to communicate the concept that the wave travels with respect to the inner system. Again, it is best to ignore the outer system. In a nested relationship, the motion of the wave is always with respect to the inner system.

Once the ray segment has reached the intercept point, it reverses direction and the reflected ray(s) travels outward from point *Q*, with one specific ray heading back to point *P* at the origin of the inner system.

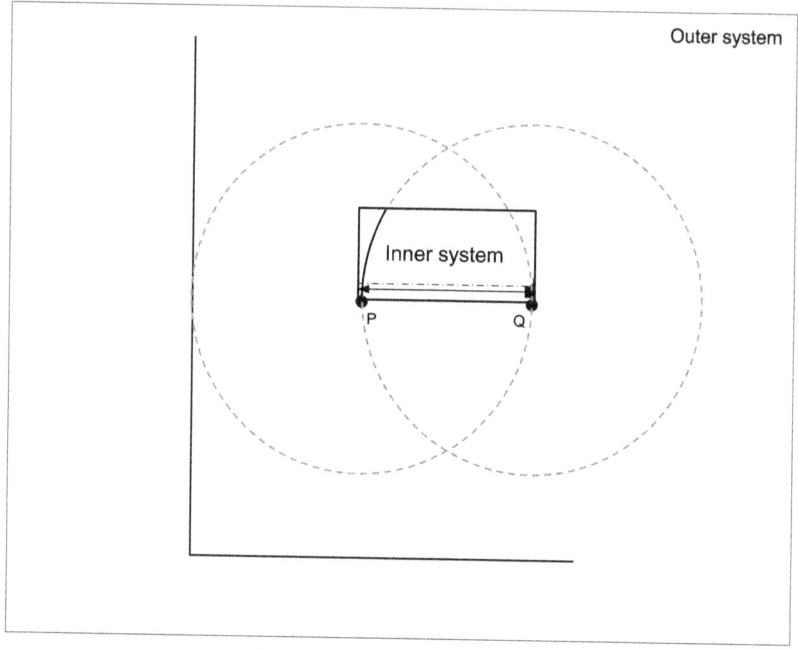

Nested system relationship
Symmetric oscillation

Figure 3–7 A nested relationship after the oscillating system has reached point P of the inner system. Because the oscillating system moves with respect to the inner system, the place in the inner system where the reflected segment intercepts point P is called the intercept point with the origin. In a nested relationship, this is simply point P, or the origin. Because the forward and reflected segments are the same length, this is called symmetric wave oscillation. Technically, the wave does not extend into the outer system. It is shown that way in the diagram so that you can see the circular (2D) or spherical (3D) waveform.

As illustrated in Figure 3–7, the forward segment, $\overrightarrow{PQ}$, and the reflected segment, $\overleftarrow{PQ}$, as measured with respect to the inner system, have the same length and require the same transit time. This motion is called **symmetric wave oscillation** and is generally described using a sinusoidal function. As long as the oscillating wave source has a positive velocity, it is able to oscillate regardless of the velocity of the inner system. In a nested relationship, there is no upper limit to the velocity of the oscillating wave source or of the inner system. From the perspective of the outer system, the velocity of the oscillating wave source *appears* to vary depending on its direction of travel. Note that the actual propagation speed is determined by the properties of the inner system's medium.

As an example of waves in a nested relationship, visualize two people in a rowboat – one in front and one in back. The rowboat (inner system) is on a lake (outer system). The person (point P) in the back puts his hand in water that has collected in the bottom of the rowboat causing a wave (oscillating system) to move in all directions, where eventually a specific ray will reach the person (point Q) at the front of the boat. Once the wave reaches the person at the front of the boat, she slaps her hand against the water in the bottom of the boat, creating a wave that propagates back toward the person at the rear of the boat. Notice that the wave's path and behavior is defined with respect to the rowboat (inner system).

In a nested relationship, the oscillating wave source moves with respect to the inner system. This means that the oscillating wave source can complete the forward segment, $\overrightarrow{PQ}$, regardless of the velocity of the inner system.

Now consider the case of a non–nested relationship for a wave where both the inner system and oscillating wave source are in motion. *In a non–nested relationship, the oscillating wave source moves with respect to the inner system, but the wave medium remains stationary since it is part of the outer system.* In this case, it is best to visualize the emission point as being associated with the outer system. The general case is illustrated in Figures 3–8 through 3–10. Assume points P and Q are on the inner system, with P at the origin and Q at the lower–right corner. As illustrated in Figure 3–8, the oscillating wave source begins with its origin coincident with the origin of the inner system and the rays move outward in all directions with respect to the outer system. The ray we will consider is the one that travels to point Q.

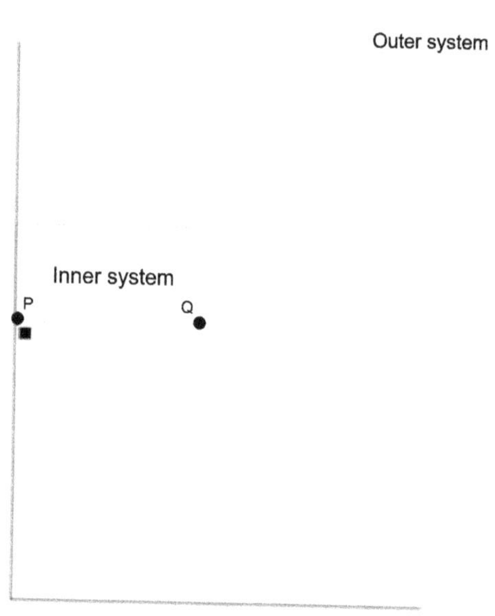

Non–nested system relationship
Asymmetric oscillation

Figure 3–8 In a non–nested relationship at time $t = 0$ the emission of the spherical wave is coincident with the origin of the inner system. The size of the spherical wave (or bubble) gets larger with the passage of time. The oscillating system moves with respect to the outer system. The wave propagates at velocity w and the inner system moves at velocity v. The wave and inner system both move with respect to the outer system.

As shown in Figure 3–9, the wave is depicted as a circle *centered at the point on the outer system that was coincident with the origin (point P) of the inner system when the wave was emitted*, the oscillating wave source travels along the path of the forward segment from the emission point to point Q at the other end of the inner system. *The oscillating wave source moves with respect to the outer system, but is oscillating between points on the inner system.* The forward segment is one ray of many that makes up the wave (which is illustrated as a circle).

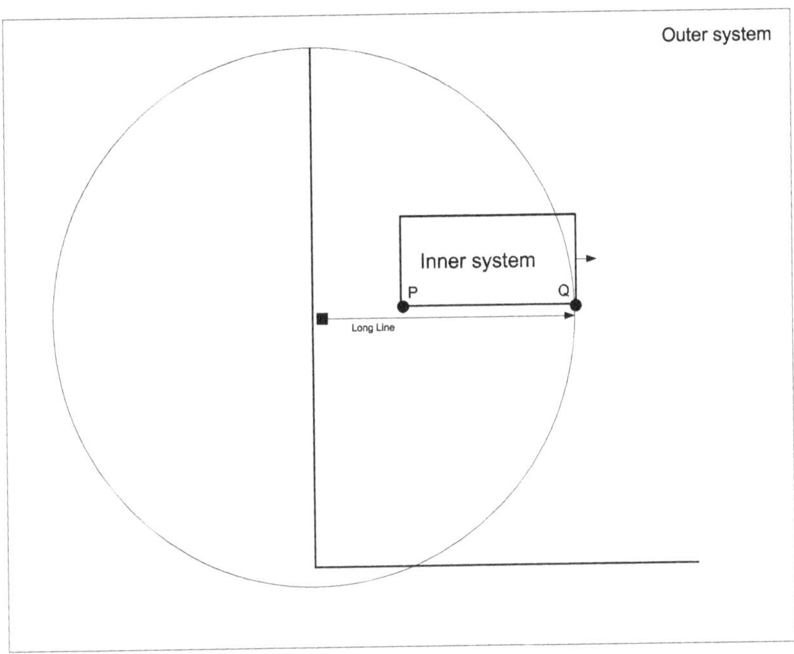

Figure 3–9 A non–nested relationship after the oscillating system has reached point Q of the inner system. The location in the outer system where the forward segment intercepts point Q of the inner system is called the intercept point. The wave propagates outward in all directions from a center on the outer system.

Once the oscillating wave has completed the forward segment, it has arrived at the intercept point. The amount of time required for the ray to travel along the forward segment, $\overrightarrow{PQ}$, is longer than when the inner system was stationary.

Since in a non–nested relationship the motion of the wave is always with respect to the outer system, the forward segment will only be completed when the velocity of the inner system is less than that of the oscillating wave source. After arriving at point Q, the oscillating wave source reverses direction and a new wave is reflected outward. We consider the specific ray of the reflected wave that travels back to the origin (point P) of the inner system.

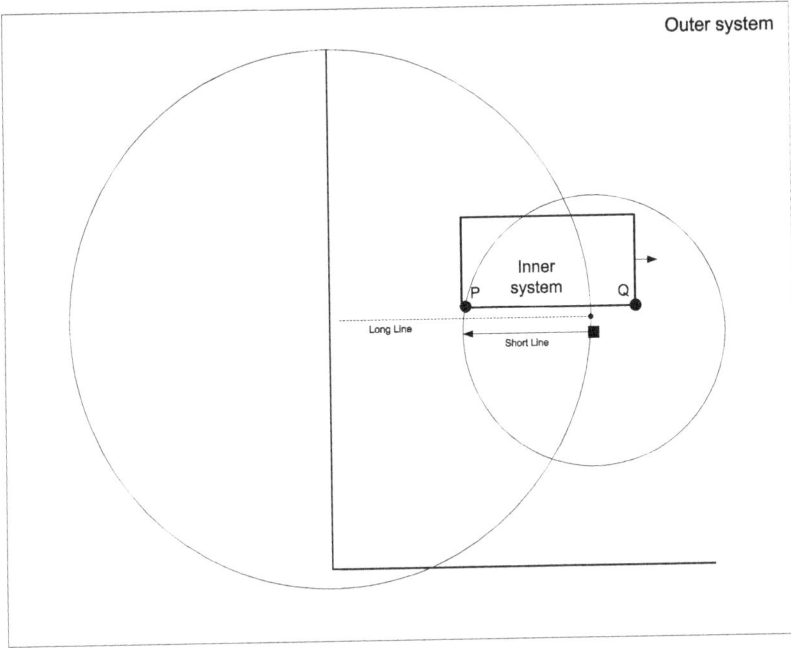

Non–nested system relationship
Asymmetric oscillation

Figure 3–10 A non–nested relationship after the reflected segment has reached point P of the inner system. Because the oscillating system moves with respect to the outer system, the place in the outer system where the reflected segment intercepts point P is called the intercept point. Because the forward and reflected segments differ in length, this is called asymmetric oscillation.

As illustrated in Figure 3–10, the reflected ray travels a shorter distance to return to the origin of the inner system. The ray is able to complete segment $\overleftarrow{PQ}$ regardless of the velocity of the inner system. In non–nested relationships with both the inner system and the oscillating system in motion, the forward and reflected segments will have different lengths. This wave motion, where the forward and reflected segments have different lengths, is called **asymmetric wave oscillation** and is generally described using a non–symmetrical sinusoidal function. The propagation speed is determined by the properties of the outer system's medium.

As an example of waves in a non–nested relationship, visualize two people in a rowboat – one in front and one in back. The rowboat (inner system) is on a lake (outer system). The person (point P) in the back puts his hand in the lake causing a wave (oscillating system) to move in all directions, where eventually a specific ray will reach the person (point Q) at the front of the boat. Once the wave reaches the person at the front of the boat, she slaps her hand against the water, creating a wave, one ray of which propagates toward the person at the rear of the boat. Notice that the wave's path and behavior is defined with respect to the lake (outer system).

Note that a non–nested relationship might be different than what is illustrated. For example, the oscillating system could be physically contained within an inner system, yet it could still behave like a non–nested relationship, because the *medium* upon which it travels is associated with the outer system.

In a non–nested relationship, the oscillating wave source moves with respect to the outer system. This means that the oscillating wave source is only able to complete the forward segment, $\overrightarrow{PQ}$, when the velocity of the inner system is less than that of the oscillating wave source. If the velocity of the inner system matches or exceeds that of the oscillating wave source, then the oscillating wave source will never reach point Q and it will not be able to reverse direction. Said simply, if the velocity of the inner system matches or exceeds that of the oscillating wave source, oscillations will not occur. When the velocity of the inner system is less than that of the oscillating wave source, then from the perspective of the inner system, both segments appear to be the same length. However, they appear to travel at different velocities and require different amounts of transit time: The wave traveling forward segment appears to move at a slower velocity and takes more time than the one traveling the reflected segment.

Summary

Chapters 2 and 3 have developed the conceptual framework for Modern Mechanics. Chapter 2 covered motion involving velocity, length, and time, showing how Modern Mechanics is able to explain the same things that are explained by classical mechanics. Chapter 3 introduced the key concepts and characteristics of three–system models in Modern Mechanics. Where Modern Mechanics differs from its second–generation cousins is in its use of three–system models to explain cyclical motion. Three–system models enable Modern Mechanics to explain cyclical motion in terms of velocity, wavelength, and frequency. Building upon its geometric transformation base, a three–system model explains oscillation using a fifth rule:

Rule 5: Cyclical behavior is explained using three–system models, where the third system, called the oscillating system, oscillates between two points that are on the inner system.

With the five rules for Modern Mechanics (four developed in Chapter 2 and one in Chapter 3), we are able to explain the behavior of moving systems in terms of length (eg, velocity, length, and time) and wavelength (eg, velocity, wavelength, and frequency). In Chapter 4, we will develop the mathematics associated with three–system models.

In a three–system model, the inner system always moves with respect to the outer system. The oscillating system moves back and forth between two points on the inner system, but moves with respect to the system on which it was placed. In a nested relationship, the oscillating system moves with respect to the inner system. In a non–nested relationship, the oscillating system moves with respect to the outer system. Notice that in two and three–system models, there is no upper limit to the velocity of a moving oscillating system. However, the velocity of the inner system is constrained if an oscillating system is required to be part of a non–nested relationship. When oscillation in a non–nested relationship is required, the velocity of the inner system must be less than that of the oscillating system, or the oscillating system will never complete the forward segment and the system will be unable to oscillate.

Three–system models are critical for explaining the observed behavior of waves in a consistent manner, regardless of whether we are talking about light, water, sound, or electromagnetic force. This unique ability to explain electromagnetic force, particle movement, and wave motion in a unified way using geometric transformations distinguishes Modern Mechanics from its first and second–generation counterparts.

Chapter 4 Essential Mathematics in Modern Mechanics

Motion in Modern Mechanics is described using the conceptual framework and terminology developed in chapters 2 and 3. Chapter 2 introduced a two–system model that explains motion involving velocity, length, and time; while Chapter 3 introduced a three–system model that explains motion involving velocity, frequency, and wavelength. In Modern Mechanics, motion is explained by using the well–understood mathematics associated with geometric transformations.

A three–system model has two characteristics that distinguish it from a two–system model. First, it consists of an outer system and *two moving systems*: the inner system and the oscillating system. Second, the oscillating system is bidirectional, meaning it moves in two directions. The equations we will develop must account for both of these characteristics. This chapter will introduce the equations and functions associated with three–system relationships in Modern Mechanics.

Averages

Rather than begin by presenting the equations and functions associated with three–system relationships, a better place to begin is to examine the *form* of an equation. The form of an equation simply refers to how an equation is written. Two equations can be equivalent – meaning they will always produce the same answer – but written differently. For example:

$$b = a + a + a + a$$

and

$$b = 4a$$

are two equivalent forms of the same equation. The ability to recognize equivalent equations in different forms is one of the most critical skills required in Modern Mechanics as well as for analyzing Einstein's work.

Fortunately, the equation that must be recognized and mastered in multiple forms is the **average**, which is also called the **arithmetic mean**. In mathematics, an average can take on different meanings. For example, it could refer to the median, mode, harmonic mean, geometric mean, or arithmetic mean. Each of these terms has a specific mathematical meaning and associated equation. In Modern Mechanics, the term average will be used interchangeably with the term arithmetic mean and is represented by the variable U_A.

You are probably familiar with the term average. The average, or arithmetic mean, of two values can be found using one of two different equations: one familiar and one unfamiliar. These two forms of the arithmetic mean equation are called the **addition mean equation** and the **subtraction mean equation**.

The familiar equation, called the addition mean equation, is:

$$U_A = \frac{U_1 + U_2}{2}$$

This equation works by finding the sum of two values, called operands, and then dividing that sum by two. The operands are represented by U_1 and U_2. For example, if you were asked to find the average of 65 and 35, you would use this equation to produce 50 as the answer, first by summing the operands and then dividing that sum by 2:

$$\begin{aligned} U_A &= \frac{65+35}{2} \\ &= \frac{100}{2} \\ &= 50 \end{aligned}$$

Now imagine for a moment that you were challenged to produce an equation that finds the average of two numbers, but with one important constraint: the equation cannot use the addition (eg, '+') operator. The new equation must use the same two operands and produce the same result, but the equation needs to be expressed in a different form.

Fortunately, the arithmetic mean can be found using a second, less–familiar equation called the subtraction mean equation. The subtraction mean equation uses the subtraction operator instead of the addition operator and is written:

$$U_A = U_1 - \frac{1}{2}(U_1 - U_2) \qquad \text{Eq. 4.1}$$

The arithmetic mean U_A for the values 35 and 65 is found using the subtraction mean equation as:

$$\begin{aligned} U_A &= 65 - \frac{1}{2}(65 - 35) \\ &= 65 - 15 \\ &= 50 \end{aligned}$$

or as:

$$\begin{aligned} U_A &= 35 - \frac{1}{2}(35 - 65) \\ &= 35 + 15 \\ &= 50 \end{aligned}$$

depending on which values are selected as U_1 and U_2. An advantage of the subtraction mean equation is that it can be used without knowing the values of the original operands in advance. This may sound counter–intuitive, but it occurs when the values of the operands are determined as part of a formula or equation.

It is easy to prove that the addition mean equation and the subtraction mean equation are equivalent and will produce the same answers. Modern Mechanics uses the subtraction mean equation, because it is easier to depict the relationship between the terms and expressions associated with the model's most important equations. One of the most important terms is called the **half–difference**, U_H, which is defined as:

$$U_H = \left| \frac{1}{2}(U_1 - U_2) \right| \qquad \text{Eq. 4.2}$$

As an example, when provided the values 65 and 35 the half–difference is:

$$U_H = \left| \frac{1}{2}(65 - 35) \right| = 15$$

Notice that the order of the operands does not matter, because the absolute value operator, which looks the same as the cardinality operator of Chapter 1, will ensure that the half–difference is positive.

When the subtraction mean equation is used, one of the operands will be equal to or greater than the other operand. Thus, we can distinguish between the two operands by relabeling the one that is greater than or equal to the value of the other as U_B and relabeling the other operand as U_S. Think of the term U_B as the bigger of the two operands and U_S as the smaller of the two operands.

Combining Equation 4.1, Equation 4.2, and the notation change just discussed to specify which operand is larger and which operand is smaller, the average is found using the equations:

$$U_A = U_B - U_H$$

and

$$U_A = U_S + U_H$$

Notice that by rearranging the equations, we can find the original values U_B and U_S. For example, if the average, U_A, and the half–difference, U_H, are known, we can find the larger value, U_B, and the smaller value, U_S, by using the following equations:

$$U_B = U_A + U_H$$

and

$$U_S = U_A - U_H$$

For example, if U_A is 50 and the half–difference U_H is 15, then U_B and U_S are:

$$\begin{aligned} U_B &= U_A + U_H \\ &= 50 + 15 \\ &= 65 \end{aligned}$$

and

$$\begin{aligned} U_S &= U_A - U_H \\ &= 50 - 15 \\ &= 35 \end{aligned}$$

Figure 4–1 illustrates the important relationship between U_H, U_B, U_S, and U_A when used in the subtraction mean equation. The ability to recognize the subtraction mean equation as an alternate way to produce the arithmetic mean is critical to understanding the Modern Mechanics equations and the relativity theory equations.

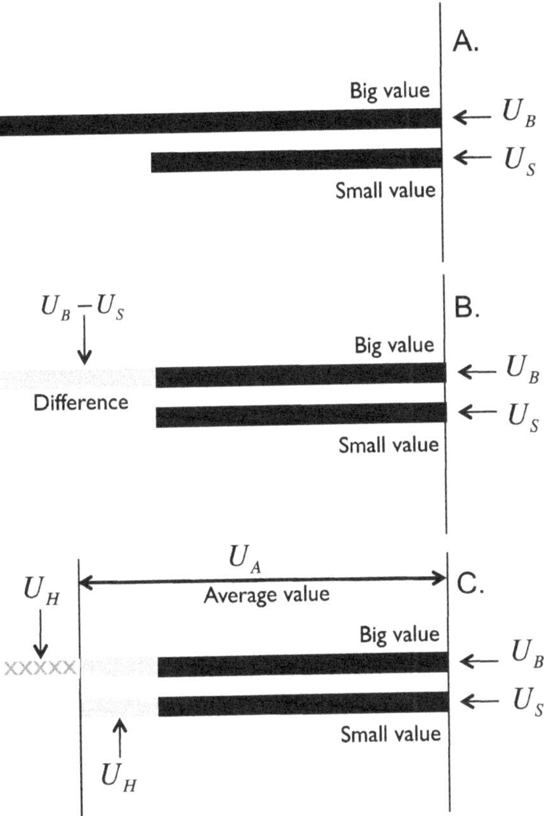

Figure 4–1 illustrates the relationship of the variables used in the subtraction mean equation. The average is found by subtracting the half-difference from the big value. The average is also found by adding the half-difference to the small value. Similarly, when given the average and half-difference, the big and small values are easily found.

The values U_S, U_B, and U_A, can be determined using a single, powerful function:

$$U(s) = \{U_A + s * U_H\} \quad \text{Fn. 4.1}$$

In Function 4.1, U is the function name and s is a function parameter that determines whether the function will return the small value U_S, the big value U_B, or the average value U_A. When the function is invoked, if s is -1, then the function returns the small value U_S; if s is 0, then the function returns the average value U_A; and if s is 1, then the function returns the big value U_B. The equations and functions we will develop for Modern Mechanics will be in this form. Note that we use { } as notation for the function body in a function definition, distinguishing it from an equation. This notation, as well as the distinction and subtle differences between functions and equations will be discussed in Chapter 5.

The ability for one function or equation to determine the large value, the small value, and the average value is extremely powerful and is what enables Modern Mechanics to serve as a unified theory. While Modern Mechanics *explicitly* defines the relationship between U_A, U_H, U_B, and U_S, these terms and equations are also *implicitly* used in other moving system models. In fact, Chapter 7 will show that U_S and U_B serve as the foundation for the Doppler equations in classical mechanics. Chapter 6 will show how U_A is the foundational equation in relativity theory and how U_H is used in Einstein's Tau function. Explaining the relationship between these terms and expressions is important, because Modern Mechanics must be able to mathematically explain the same experiments and observations as its first and second–generation counterparts.

Essential Equations

The mathematical relationship between systems is described using three types of equations: **position**, **distance**, and **duration**.

Duration is an expression of time. While the meaning of time will be discussed in Chapter 8, in this chapter we will use the terms *duration* and *time* interchangeably. Duration equations describe the amount of time required for a system to travel from a starting position to an ending position with respect to its outer system. Similar to the length equation introduced in Chapter 2, the duration equation defines a relationship between *time*, *velocity* and *distance*. Time t_w is found as:

$$t_w = \frac{d}{w} \qquad \text{Eq. 4.3}$$

where d is a distance, and w is a velocity. Here we use w to represent velocity rather than v, which is traditionally used. This choice in variables will facilitate the development of the remaining equations for three–system models in the remainder of this chapter.

Distance equations describe the distance between two points on a system. They also describe the distance that a system moves with respect to its outer system. While often thought of as a length, a distance can also describe the amount of effort required in moving from a starting position to an ending position. This relationship, described as energy, will be discussed in Chapter 7. Distance equations are useful in describing motion, which is especially important when working with velocity, wavelength and frequency.

The **static distance** between two points is determined using the equation:

$$\sqrt{(x-x_0)^2+(y-y_0)^2+(z-z_0)^2}=d$$

where d represents the distance between two points: (x,y,z) and (x_0,y_0,z_0). To simplify the development of the Modern Mechanics equations, we assume the origin as one of the points, simplifying the static distance equation as:

$$\sqrt{x^2 + y^2 + z^2} = d \qquad \text{Eq. 4.4}$$

In formal notation, this equation would be written as:

$$\exists(x,y,z) : \sqrt{x^2 + y^2 + z^2} = d$$

where $\exists(x,y,z)$ means "there is a point such that" the equation is satisfied.

The equation for a **dynamic distance** begins as a variant of the duration Equation 4.3 and is written as:

$$d = wt_w \qquad \text{Eq. 4.5}$$

where w is the velocity of the moving system and t_w is the amount of time that it has been moving. By combining equations 4.4 and 4.5, the dynamic distance equation is defined in terms of time and velocity to reach the specific point (x, y, z) as:

$$\sqrt{x^2 + y^2 + z^2} = wt_w \qquad \text{Eq. 4.6}$$

This dynamic distance equation expresses the distance between the origin and a specific point in terms of velocity and time. Dynamic distance equations are useful to describe the length traversed by a system. As will be developed in this chapter, they are particularly useful to describe the behavior of an oscillating system in a three–system model.

Generally, equations 4.4, 4.5, and 4.6, are used to represent distances to a specific point (x, y, z). Notice that velocity multiplied by duration is a distance; it is not a position. As illustrated in Figure 4–2, a distance and a position are different, because a distance does not need to correspond to any of the coordinates that make up a position. Notice that the distance traveled, d, does not need to correspond to any of the coordinates that define the point (x, y, z). However, if the point is on one of the axes, then the coordinate value on that axis will have the same value as the distance d.

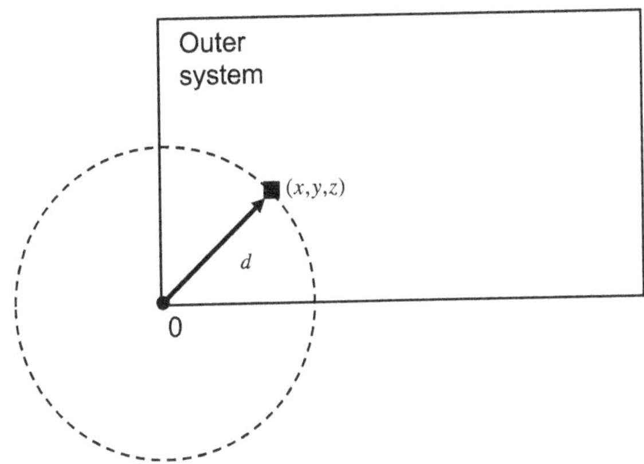

Figure 4–2 The distance of a single ray to the point (x, y, z). Notice that the distance could apply to any point on the sphere, but in the case of the distance equation applies to a specific point. Note that a circle is shown in the illustration rather than a sphere.

In Modern Mechanics, the dynamic distance equation, Equation 4.6, is used to find length equations associated with the motion of an oscillating system. These length equations are called the **forward intercept length**, the **reflected intercept length**, and the **average intercept length**.

We are also able to find time equations associated with the motion of an oscillating system. These time equations are called the **forward intercept time**, the **reflected intercept time**, and the **average intercept time** of an oscillating system. These time equations are formed by rearranging Equation 4.6 to arrive at:

$$\frac{\sqrt{x^2+y^2+z^2}}{w}=t_w \qquad \text{Eq. 4.7}$$

Surface equations describe geometric shapes. *A surface is the set of all points* that describe the surface of a shape. They differ from position or distance equations because *surface equations describe the simultaneous behavior of a collection of points*, rather than the behavior of a specific, individual point. One of the most important shapes used in Modern Mechanics is the sphere or spherical wave. The equation for a sphere is:

$$x^2+y^2+z^2=R^2 \qquad \text{Eq. 4.8}$$

where R *is the radius for every point of the surface.* Unlike equations 4.4, 4.5, 4.6, and 4.7, which are associated with one point, Equation 4.8 simultaneously describes all of the points that make the equation true for a given radius. All points on the surface of the sphere are referred to as (x,y,z). Admittedly, the distinction between Equation 4.8 and the preceding equations is hard to describe without the use of formal math notion to represent "there exists a point such that" and "all points such that." What is important is to recognize that, when describing a surface, the equation represents all possible combinations of x, y, and z that make that equation true. In formal notation, Equation 4.8 could be written as:

$$\forall(x,y,z): x^2+y^2+z^2=R^2$$

where $\forall(x,y,z)$ means that we must consider the collection of "all (x,y,z) points" that satisfy the equation. In Modern Mechanics we distinguish between d, which is used to represent the distance to a specific individual point, from R, which is used to represent the distance to a collection of points.

Similar to a distance, a radius can be static or dynamic. Equation 4.8 describes a static sphere, which means that the radius does not vary with time. Alternatively, the radius can be dynamic, where the size of the sphere or spherical wave varies with time. When the radius describes a dynamic sphere, the radius is determined by the velocity, w, of the spherical wave and the amount of time that has elapsed since the wave was emitted, t_w, as:

$$R = wt_w \qquad \text{Eq. 4.9}$$

Distance equations are used differently on points and surfaces. Equation 4.5 describes a *distance to a specific point,* while Equation 4.9 describes the *radius to a collection of points.* This subtlety between a distance and a radius is easily overlooked.

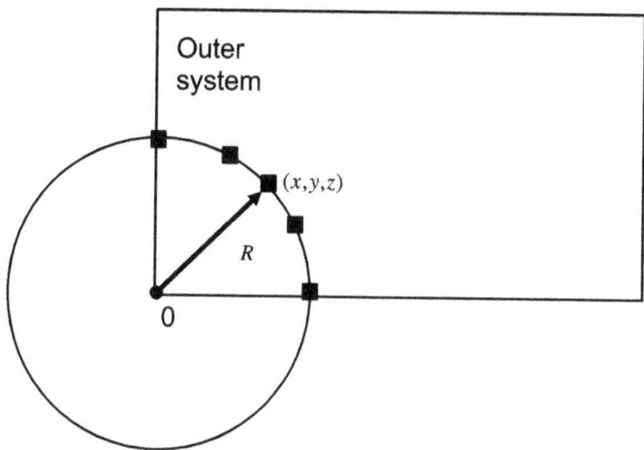

Figure 4–3 The distance of every ray. The radius describes every point on a surface, though only one of its points is highlighted with an accompanying vector. In this case, the equation describes every possible combination of (x, y, z) that make the equation true. The radius of a sphere represents a surface comprised of an infinite number of points. Note that a circle is shown in the illustration rather than a sphere.

Combining equations 4.8 and 4.9, the equation for a spherical wave, which is also called a dynamic sphere, is:

$$x^2 + y^2 + z^2 = w^2 t_w^{\ 2} \qquad \text{Eq. 4.10}$$

The amount of care that must be taken to recognize that equations 4.9 and 4.10 apply to a set of points and not to an individual point cannot be understated. Equations 4.8, 4.9, and 4.10 solve for all values of x, y, and z that make the equations true. Notice that equations 4.6 and 4.10 could be written identically if Equation 4.6 was squared. However, *Equation 4.6 is solved for one point,* $\exists(x, y, z)$*, that makes the distance true*, while *Equation 4.10 solves for every combination of points,* $\forall(x, y, z)$*, that makes the equation true.* This distinction can only be detected if stated textually, through the context of the equations, or when formal math notation is used.

Position equations describe the location of a system with respect to its outer system. Positional changes of a system are performed using geometric transformations. Modern Mechanics uses the translation transformation to change the position of a system, where its new position is found by adding a distance to its starting position. The general equation describing a change in position along the x axis is:

$$x' = x + d \qquad \text{Eq. 4.11}$$

where x is the original position, d is the static distance that the system will be moved, and x' is the new position. Note that the choice of variables to represent the starting and ending positions is arbitrary. However, since classical mechanics traditionally uses x as the starting position and x' as the ending position, Modern Mechanics will continue with this nomenclature.

Because Modern Mechanics must describe motion in terms of velocity and time, we combine equations 4.5 and 4.11 to produce:

$$x' = x + vt_v \qquad \text{Eq. 4.12}$$

Notice that we have renamed the variables used in Equation 4.5 and now use v to describe velocity rather than w, and t_v to describe duration rather than t_w. This use of variable names will facilitate our discussion of three–system models, where w and t_w are associated with the time and velocity of the oscillating system, and v and t_v are associated with the time and velocity of the inner system.

The position equation is useful to describe the motion of a system in terms of velocity, length, and time. Figure 4–4 illustrates the positional change of an inner system with respect to an outer system by using the translation transformation. For simplicity, begin with the origin of the inner system coincident with the origin of the outer system. The length of the inner system along the x axis is x. When both origins are coincident, a point at x in the inner system is also at location x of the outer system. The inner system moves d units to the right using the translation equation so that the point x in the inner system arrives at a new point, x', on the outer system.

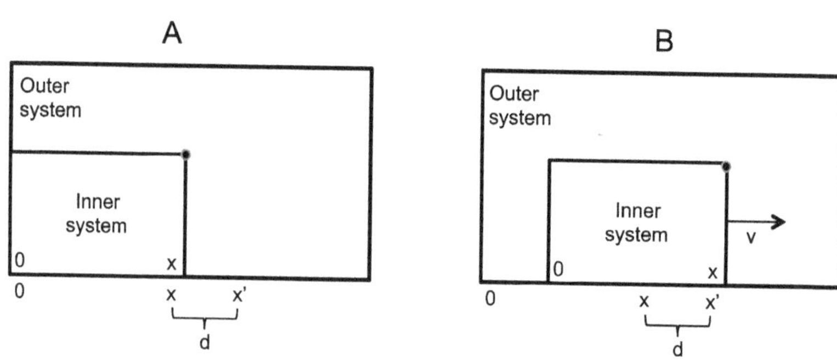

Figure 4–4 A two-system model where the inner system moves with respect to the outer system. In (A), the inner system is in its original position. In (B), the inner system has moved to the right using translation. The amount of movement is determined by Equation 4.11 when the motion is static or by Equation 4.12 when the motion is dynamic.

While Equation 4.12 explains motion along the x axis with respect to velocity and time, the position equations in Modern Mechanics must explain motion in every possible direction. Motion in any direction of an inner system is defined by using a velocity vector $\vec{v}$, which has three components, $< v_x, v_y, v_z >$, resulting in the following positional equations:

$$\begin{aligned} x' &= x + v_x t_V \\ y' &= y + v_y t_V \\ z' &= z + v_z t_V \end{aligned} \qquad \text{Eqs. 4.13}$$

In Equation 4.13, the beginning position is (x, y, z) and the ending position is (x', y', z').

Vectors are extremely useful for describing the directional information of a system. A key advantage of vectors is that they can describe motion and velocity in any direction with three numbers. While vectors may not be familiar to all readers, the concept is similar to a human color palette, where we can create an infinite number of colors by mixing various amounts of the three primary colors. Similarly, vectors allow us to describe an infinite number of directions and velocities in three–dimensional space using only three numbers.

Because it is easy to confuse a distance with a position, care must be taken when using equations 4.5 and 4.12. Equation 4.5 produces a distance, while Equation 4.12 produces a position. Consider the case where $x = 0$:

$$\begin{aligned} x' &= x + vt_v \\ x' &= 0 + vt_v \\ x' &= vt_v \end{aligned} \qquad \text{Eqs. 4.14}$$

which is the equation for a *position*. However, because it is written in a simplified form, it is easily confused for a distance. In this equation, vt_v is a distance, but $0 + vt_v$ is a position, which makes x' a position. This distinction is subtle but important, because the mistreatment of a distance as a position, or vice versa, can lead to mathematical mistakes.

Intercept Equations

The interactions of a three–system model are best described in terms of nine equations. These equations are the forward intercept length, reflected intercept length, average intercept length, forward intercept time, reflected intercept time, average intercept time, inner system position, oscillating system position, and intercept position.

In a three–system model, intercept lengths, intercept times, and positions define the relationship – in terms of space, time, and distance – between an oscillating system and the inner system. A three–system relationship can be nested or non–nested. We will first develop the mathematics for a non–nested relationship, because the mathematics associated with a nested relationship are a simplified form of the former.

Non–Nested Intercept Equations

A three–system model requires intercept equations that associate an inner system and an oscillating system. We will first develop the equations for the forward intercept length and time.

The inner system begins with its origin coincident with the origin of the outer system. The oscillating system also starts at the origin. The inner system moves with velocity v and the oscillating system moves at velocity w. The oscillating system travels with respect to the outer system toward the point (x, y, z) on the inner system.

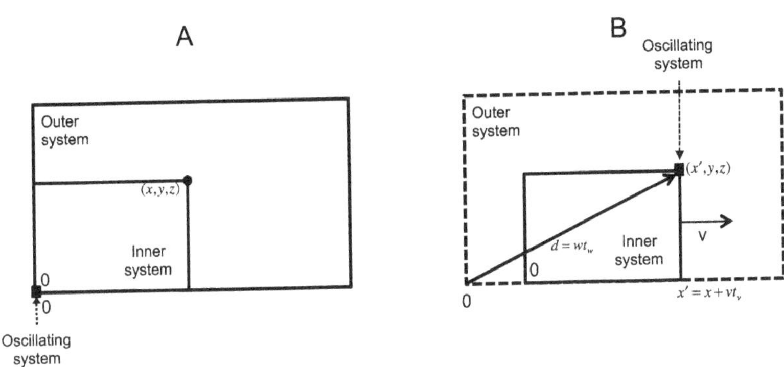

Figure 4–5 In a three–system, non–nested relationship, the intercept point is defined as the point on the outer system where the oscillating system intercepts the point (x, y, z) on the inner system, which is also in translatory motion. In this case, the inner system and the oscillating system both move with respect to the outer system, thus the intercept point on the outer system is (x',y,z). The forward intercept length, or the distance traveled by the oscillating system, is found using the distance equations. The intercept position is found using Equation 4.13.

As stated earlier, w represents the velocity of the oscillating system and t_w represents the amount of time it has been in motion; v represents the velocity of the inner system and t_v represents the time it has been in motion. The intercept position occurs when the position of the oscillating system is coincident with the point (x, y, z) of the inner system. Since both the oscillating system and the point (x, y, z) of the inner system move with respect to the outer system, the intercept position is at (x', y, z) on the outer system.

We have to account for the inner system's motion in any direction, because it *is not required* to move solely along the x axis. By using Equation 4.13 to govern the movement of the inner system, the intercept point will be located at (x', y', z') with respect to the outer system.

The amount of time, t_w, that the oscillating system has been in transit from the origin to the forward intercept position is the same as the amount of time that the inner system has been moving, t_v. This is called the **forward intercept time** and is denoted t_F. By using the forward intercept position (x',y',z') and Equation 4.6, we compute the forward intercept length, wt_F, as:

$$\sqrt{x'^2+y'^2+z'^2} = wt_F \qquad \text{Eq. 4.15}$$

where t_F is the amount of time for the oscillating and inner systems to arrive at the forward intercept position. Equation 4.15 is one of the most important equations in Modern Mechanics. Once solved, it defines the forward intercept position (x',y',z'), the forward intercept time, t_F, and the forward intercept length, wt_F. However, this equation cannot be solved directly, since all of the variables, except the velocity of the oscillating system, w, are unknown.

Fortunately, Equation 4.15 can be expressed in terms of x, y, and z by combining it with Equation 4.13 to produce:

$$\sqrt{(x+v_xt_F)^2+(y+v_yt_F)^2+(z+v_zt_F)^2} = wt_F \qquad \text{Eq. 4.16}$$

Since the point (x,y,z) on the inner system is given and the velocity $\vec{v}$ of the inner system is given, this equation can now be solved in terms of the forward intercept time t_F. However, since the term t_F appears on both sides of the equation, including within the radical, it cannot be easily solved using algebraic substitution. Instead, this equation is solved using the quadratic equation as:

$$t_F =$$

$$\frac{\frac{1}{2}\sqrt{(-2xv_x-2yv_y-2zv_z)^2-4(-x^2-y^2-z^2)(w^2-v_x^2-v_y^2-v_z^2)}+(xv_x+yv_y+zv_z)}{w^2-v_x^2-v_y^2-v_z^2}$$

Eq. 4.17

This equation defines the amount of time for an oscillating system to move from the origin to intercept the point (x, y, z) on the inner system. Since both the inner system and the oscillating system move with respect to the outer system, this interception occurs at the point at (x', y', z') on the outer system.

Once an oscillating system arrives at the forward intercept point, it is reflected and travels back toward the origin of the inner system. The equation for the **reflected intercept time** is solved in a similar manner as:

$$t_R = \frac{\frac{1}{2}\sqrt{(-2xv_x - 2yv_y - 2zv_z)^2 - 4(-x^2 - y^2 - z^2)(w^2 - v_x^2 - v_y^2 - v_z^2)} - (xv_x + yv_y + zv_z)}{w^2 - v_x^2 - v_y^2 - v_z^2}$$

Eq. 4.18

where t_R is the reflected intercept time.

The **average intercept time**, t_A, is the average of the forward intercept and reflected intercept times and easily found by using equations 4.17 and 4.18 as:

$$t_A = \frac{\frac{1}{2}\sqrt{(-2xv_x - 2yv_y - 2zv_z)^2 - 4(-x^2 - y^2 - z^2)(w^2 - v_x^2 - v_y^2 - v_z^2)}}{w^2 - v_x^2 - v_y^2 - v_z^2}$$

Eq. 4.19

Notice that the average intercept time is not known in advance, but is instead found as part of a calculation.

Using the subtraction mean equation, the **intercept time half–difference**, t_H, is:

$$t_H = \frac{xv_x + yv_y + zv_z}{w^2 - v_x^2 - v_y^2 - v_z^2} \qquad \text{Eq. 4.20}$$

Figure 4–6 defines the relationship between the forward intercept time, reflected intercept time, average intercept time, and intercept time half–difference.

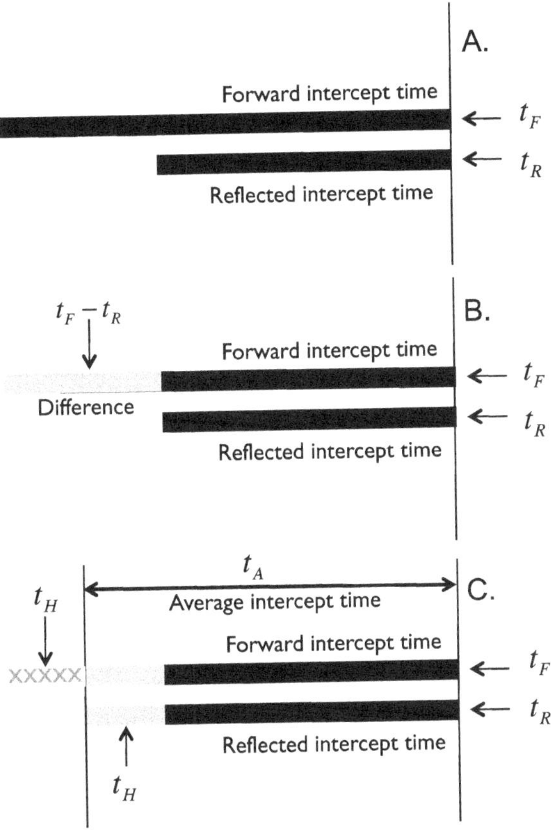

Figure 4–6 The subtraction mean equation illustrates the relationship between the forward intercept time, the reflected intercept time, the average intercept time, and the intercept time half–difference. The average intercept time is found by subtracting the intercept time half–difference from the forward intercept time. The average intercept time is also found by adding the intercept time half–difference to the reflected intercept time.

By combining equations 4.19 and 4.20 with Function 4.1, we develop a powerful function that returns the forward, reflected, and average intercept times. This time function, $T()$, is defined as:

$$T(s) = \{t_A + s * t_H\} \qquad \text{Fn. 4.2}$$

This function returns the reflected intercept time when $s = -1$, the average intercept time when $s = 0$, and the forward intercept time when $s = 1$. When parameters for the function variables and the expressions for the average intercept time and intercept time half–difference are replaced in the function, the function is defined as:

$$T(x,y,z,\vec{v},s,w) = \left\{ \frac{\frac{1}{2}\sqrt{(-2xv_x - 2yv_y - 2zv_z)^2 - 4(-x^2 - y^2 - z^2)(w^2 - v_x^2 - v_y^2 - v_z^2)} + s(xv_x + yv_y + zv_z)}{w^2 - v_x^2 - v_y^2 - v_z^2} \right\}$$

Fn. 4.3

where $\vec{v}$ is the velocity vector and s is the selector. In Modern Mechanics, Function 4.3 is the **intercept time function** and will produce the forward, reflected, and average intercept times for a non–nested relationship.

As a simplifying assumption, *consider the specific case where the translatory motion of the inner system occurs only along the x axis of the outer system*, such that $v_y = 0$ and $v_z = 0$. When we only consider the v_x component of the velocity vector, the intercept time function simplifies as:

$$T(x,y,z,\vec{v},s,w) = \left\{ \frac{\frac{1}{2}\sqrt{(-2xv_x)^2 - 4(-x^2 - y^2 - z^2)(w^2 - v_x^2)} + s(xv_x)}{w^2 - v_x^2} \right\}$$

which is further simplified as:

$$T(x,y,z,v_x,s,w) = \left\{ \frac{\sqrt{w^2x^2 + w^2y^2 + w^2z^2 - y^2v_x^2 - z^2v_x^2} + s(xv_x)}{w^2 - v_x^2} \right\}$$

Fn. 4.4

Notice that since the inner system moves in translatory motion along the x axis, only the x axis velocity component v_x is required by the function.

Since Equation 4.5 defines a velocity multiplied by time as a distance, we can combine it with Function 4.3 to define an **intercept length function** as:

$$L(x,y,z,\vec{v},s,w) = \{w * T(x,y,z,\vec{v},s,w)\} \qquad \text{Fn. 4.5}$$

The intercept length function returns the reflected intercept length when $s = -1$, the average intercept length when $s = 0$, and the forward intercept length when $s = 1$. Similar to the intercept *time* function, the generalized intercept *length* function is simplified as:

$$L(x,y,z,v_x,s,w) = \{w * T(x,y,z,v_x,s,w)\} \qquad \text{Fn. 4.6}$$

if we consider only the v_x component of the velocity vector to show the inner system's translatory motion solely along the x axis.

We now determine the forward intercept position of the point (x, y, z) on the inner system with respect to the outer system by using the translation equation. Because T_F is $T(x,y,z,\vec{v},1,w)$, the resulting **forward intercept position** transformations are:

$$\begin{aligned} x' &= x + v_x T(x,y,z,\vec{v},1,w) \\ y' &= y + v_y T(x,y,z,\vec{v},1,w) \\ z' &= z + v_z T(x,y,z,\vec{v},1,w) \end{aligned} \qquad \text{Eq. 4.21}$$

The reflected intercept position, with respect to the outer system is found by adding the reflected intercept length, $\vec{v}T_R$, to the origin. Because T_R is $T(x,y,z,\vec{v},-1,w)$, the resulting **reflected intercept position** transformations are:

$$x' = 0 + v_x T(x,y,z,\vec{v},-1,w)$$
$$y' = 0 + v_y T(x,y,z,\vec{v},-1,w) \quad \text{Eq. 4.22}$$
$$z' = 0 + v_z T(x,y,z,\vec{v},-1,w)$$

assuming that the origins of the outer and inner systems are initially coincident and the oscillating system started at (x, y, z) rather than the origin.

We can also find the average intercept position that, as an average, is not a true intercept position. When both the inner system and the oscillating system are in motion, it is not always guaranteed that the average intercept time and average intercept length will coincide with either intercept point. As will be discussed in Chapter 6, this is a significant distinction between Modern Mechanics and relativity theory: Einstein asserts that this average intercept represents a forward or reflected intercept. One of the goals of Einstein's spherical wave proof is to show this relationship.

Now, consider three specific examples to illustrate the use of the intercept time function, the intercept length function, and the intercept position equations. Figure 4–7 illustrates three interactions where an oscillating system travels between the origin of an inner system to three specific points on the inner system.

- In Figure 4–7(A), the oscillating system moves back and forth between the origin of the inner system and a point on the y axis, located at position $(0, y, 0)$ on the inner system.

- In Figure 4–7(B), the oscillating system moves back and forth between the origin of the inner system and a point on the x axis, located at position $(x, 0,0)$ on the inner system.

- In Figure 4–7(C), representing the general case, the oscillating system moves back and forth between the origin of the inner system and a point at position (x, y, z) on the inner system.

Each example will be described in terms of the forward, reflected, and average intercept lengths, times, and positions. These examples will illustrate how time and length functions and equations are used in Modern Mechanics, how they are related to classical mechanics' Doppler equations, and will be used in Chapter 6 to explain relativity theory.

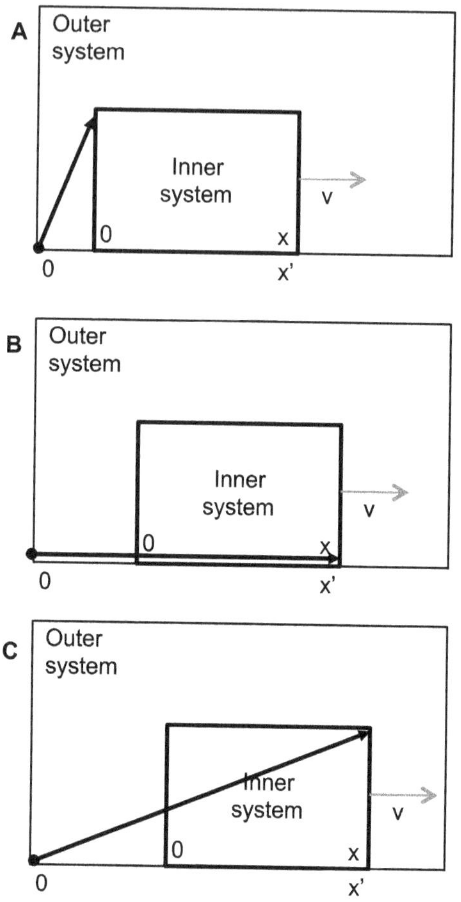

Figure 4–7 The inner and oscillating systems move with respect to the outer system using the translation transformation. Each example is a non–nested relationship where the velocity of the inner system, v, is less than the velocity of the oscillating system, w. In (A), the oscillating system moves from the origin to a point $(0, y, 0)$ at the top of the inner system. In (B), the oscillating system moves to a point $(x, 0, 0)$ at the end of the inner system. In (C), the oscillating system moves to a point (x, y, z) at an arbitrary position on the inner system. Note that (C) is the most general case. The z axis interaction is discussed in the text, but not illustrated.

In Figure 4–7(A), the oscillating system moves back and forth along the y axis of the inner system, between the origin and $(0, y, 0)$ of the inner system. By invoking the *forward intercept time* function, the amount of time for the oscillating system to move from the origin to the point is found as:

$$\tau_{yF} = T(0, y, 0, v, 1, w) \equiv \frac{y}{\sqrt{w^2 - v^2}}$$

In this function invocation, τ_{yF} is the forward intercept time instance variable to a point on the y axis, $T(0, y, 0, v, 1, w)$ is the function invocation, and $\frac{y}{\sqrt{w^2 - v^2}}$ is the instantiated equation.

We use $'\equiv'$ to denote the equivalence between a function invocation and its instantiated equation, enabling us to use the equation instead of the function invocation. The terminology, nuances, and subtleties of functions and equations are discussed in Chapter 5.

The *reflected intercept time* from a point on the y axis to the origin, τ_{yR}, is found as:

$$\tau_{yR} = T(0, y, 0, v, -1, w) \equiv \frac{y}{\sqrt{w^2 - v^2}}$$

In the case of the inner system's translatory motion along the x axis, the forward and reflected intercept times are the same for oscillation between the origin and a point on the y axis. As a result, the *average intercept time*, τ_{yA}, is:

$$\tau_{yA} = T(0, y, 0, v, 0, w) \equiv \frac{y}{\sqrt{w^2 - v^2}} \qquad \text{Eq. 4.23}$$

With the intercept times known, we can easily find the related intercept lengths. The forward intercept length ξ_{yF}, for the distance the oscillating system travels from the origin to a point on the y axis of the inner system is found by multiplying the forward intercept time by the velocity of the oscillating system, w. When invoking the forward intercept length function to perform this multiplication, we find the *forward intercept length*, ξ_{yF}, as:

$$\xi_{yF} = L(0,y,0,v,1,w) \equiv \frac{wy}{\sqrt{w^2 - v^2}} = \frac{y}{\sqrt{1 - \frac{v^2}{w^2}}}$$

Similarly, the *reflected intercept length*, ξ_{yR}, is:

$$\xi_{yR} = L(0,y,0,v,-1,w) \equiv \frac{wy}{\sqrt{w^2 - v^2}} = \frac{y}{\sqrt{1 - \frac{v^2}{w^2}}}$$

The *average intercept length*, ξ_{yA}, for the oscillating system is:

$$\xi_{yA} = L(0,y,0,v,0,w) \equiv \frac{wy}{\sqrt{w^2 - v^2}} = \frac{y}{\sqrt{1 - \frac{v^2}{w^2}}} \qquad \text{Eq. 4.24}$$

The *forward*, *reflected*, and *average intercept positions* are found using Equation 4.13, where time, t_v, is replaced by the intercept times found above.

While not illustrated, the intercept time, distances and positions for a point at $(0,0,z)$ along the z axis are found in a similar manner as for the y axis. The *forward intercept time* τ_{zF} to a point on the z axis, is found as:

$$\tau_{zF} = T(0,0,z,v,1,w) \equiv \frac{z}{\sqrt{w^2 - v^2}}$$

The *reflected intercept time*, τ_{zR}, is:

$$\tau_{zR} = T(0,0,z,v,-1,w) \equiv \frac{z}{\sqrt{w^2 - v^2}}$$

Because the forward and reflected intercept times are the same, we know that the average intercept time between the origin and a point on the z axis will be the same as well. As a result, the *average intercept time*, τ_{zA}, is:

$$\tau_{zA} = T(0,0,z,v,0,w) \equiv \frac{z}{\sqrt{w^2 - v^2}} \qquad \text{Eq. 4.25}$$

As was the case for the oscillating system traveling to a point along the y axis, we can now use the intercept times to determine the intercept lengths by multiplying them by the velocity of the oscillating system, w, to find the *forward intercept length* ξ_{zF} as:

$$\xi_{zF} = L(0,0,z,v,1,w) \equiv \frac{wz}{\sqrt{w^2 - v^2}} = \frac{z}{\sqrt{1 - \frac{v^2}{w^2}}}$$

The *reflected intercept length*, ξ_{zR}, is:

$$\xi_{zR} = L(0,0,z,v,-1,w) \equiv \frac{wz}{\sqrt{w^2 - v^2}} = \frac{z}{\sqrt{1 - \frac{v^2}{w^2}}}$$

The *average intercept length*, ξ_{zA}, for the oscillating system is:

$$\xi_{zA} = L(0,0,z,v,0,w) \equiv \frac{wz}{\sqrt{w^2 - v^2}} = \frac{z}{\sqrt{1 - \frac{v^2}{w^2}}} \qquad \text{Eq. 4.26}$$

The *forward, reflected,* and *average intercept positions* are found using Equation 4.13, where time, t_v, is replaced by the intercept times found above.

In Figure 4–7(B), the oscillating system moves with respect to the outer system between the origin and the point $(x, 0, 0)$ on the inner system. Using the intercept time function, the *forward intercept time* τ_{xF} to a point on the x axis is found as:

$$\tau_{xF} = T(x,0,0,v,1,w) \equiv \frac{x}{w-v}$$

The *reflected intercept time,* τ_{xR}, is found as:

$$\tau_{xR} = T(x,0,0,v,-1,w) \equiv \frac{x}{w+v}$$

As a result, the *average intercept time,* τ_{xA}, is:

$$\tau_{xA} = T(x,0,0,v,0,w) \equiv \frac{wx}{w^2 - v^2} \quad \text{Eq. 4.27}$$

The relationship of the values is depicted in Figure 4–8, with the intercept time function based on the subtraction mean equation.

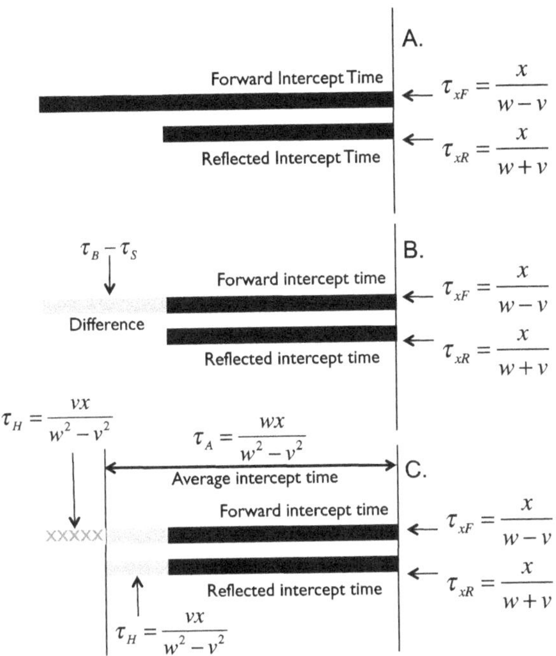

Figure 4–8 The relationship between the forward, reflected, and average times for an oscillating system along the x axis using the subtraction mean equation. The average intercept time is found by subtracting the intercept time half-difference from the forward intercept time. The average intercept time is also found by adding the intercept time half-difference to the reflected intercept time.

With the intercept times now known, we can find the intercept lengths by multiplying them by the velocity of the oscillating system, w. Using the intercept length function, we determine the *forward intercept length*, ξ_{xF}, for the oscillating system traveling along the x axis to the point $(x, 0, 0)$ on the inner system as:

$$\xi_{xF} = L(x,0,0,v,1,w) \equiv \frac{wx}{w-v} = \frac{x}{1-\frac{v}{w}} \quad \text{Eq. 4.28}$$

The *reflected intercept length*, ξ_{xR}, is:

$$\xi_{xR} = L(x,0,0,v,-1,w) \equiv \frac{wx}{w+v} = \frac{x}{1+\frac{v}{w}} \qquad \text{Eq. 4.29}$$

The *average intercept length*, ξ_{xA}, is:

$$\xi_{xA} = L(x,0,0,v,0,w) \equiv \frac{w^2 x}{w^2 - v^2} = \frac{x}{1-\frac{v^2}{w^2}} \qquad \text{Eq. 4.30}$$

The *forward*, *reflected*, and *average intercept positions* are found using Equation 4.13, where time, t_v, is replaced by the intercept times found above.

Notice that the solution for the general case in Figure 4–7(C) is readily solved using functions 4.3, 4.5, and Equation 4.13.

Some readers may recognize equations 4.28 and 4.29 – the forward and reflected intercept lengths along the x axis – as the Doppler shift equations. Additionally, readers familiar with relativity theory may recognize equations 4.24, 4.26, and 4.30 – the average intercept lengths – as Einstein's coordinate equations prior to his final substitution of *"x' with its value."* It is important to reiterate that intercept lengths do not represent positions in space. Each intercept length, ξ_{xA}, ξ_{yA}, and ξ_{zA}, represents the average of the two segments of an oscillation and it is incorrect to treat each as a coordinate. The collection of distances to three different and distinct points do not represent a transformed point $(\xi_{xA}, \xi_{yA}, \xi_{zA})$. Each length, ξ_{xA}, ξ_{yA}, and ξ_{zA}, is the average intercept distance to a point along each axis of the inner system and they do not form a spatial coordinate. We will examine Einstein's incorrect use of average intercept lengths to represent a point in space when we review his relativity derivation in Chapter 6.

While we have developed the equations for a specific ray, the intercept length can be used in conjunction with the surface equation to define the behavior of a spherical wave. We follow the same approach as discussed for a distance, only we replace d with R, because a spherical wave is a collection of rays.

As defined in Chapter 3, the intercept times and intercept lengths are only valid in a non–nested relationship when the velocity of the inner system is less than that of the oscillating system. When this velocity is exceeded, oscillations do not occur.

Nested Intercept Equations

Now that we have developed the equations for a non–nested relationship, we can examine the case of a nested relationship. In a nested relationship, the inner system takes on a dual role. It is an inner system and it also serves as the outer system for the oscillating system. In a nested relationship, the same intercept time and intercept length functions apply. However, because the inner system appears stationary with respect to the oscillating system, the velocity value would be **0**. The position of the inner system, however, is still found using the intercept position equations, because it moves with respect to the outer system.

Summary

Modern Mechanics uses geometric transformations to explain motion involving two systems. Two–system models effectively explain non–oscillating motion in terms of length, velocity, and time. The oscillating motion of a three–system model is explained using two key functions for time and length, and one key equation for position. The intercept time function will return the forward, reflected, and average intercept times. The intercept length function will return the forward, reflected, and average intercept lengths. The intercept time can be used with the intercept position equation to determine the position of the inner system at specific times. As a word of caution: *At the average intercept time (and distance), the oscillating system is not guaranteed to be at the intercept point.* The position and distance equations are useful for explaining the motion of particles, while the surface and distance equations are useful for explaining the motion of waves.

Chapter 5 The Differences Between Functions and Equations

In chapters 2 through 4, we defined the essential concepts and mathematics associated with Modern Mechanics. We will use this foundation to explain Einstein's work. As you will learn in Chapter 6, one of the most important statements in Einstein's 1905 paper that defines special relativity, *Zur Elektrodynamik bewegter Körper (On the Electrodynamics of Moving Bodies)*, begins with: "*Since τ is a linear function.*" Mathematically, functions share several similarities with equations. They also differ in some rather subtle, but extremely important ways. To effectively explain Einstein's work and identify his mistakes, we must first explain the differences between functions and equations.

Mathematics is largely based on developing and using three concepts: **operations and formulas**, **rules**, and **abstraction**. *Operations and formulas* are the things you can do, such as add, subtract, multiply, and divide. As one learns more about math, the number of operations and formulas you are introduced to and able to use increases. For example, trigonometry introduces sine, cosine, and tangent. Similarly, calculus introduces limits, derivatives, and integrals.

Rules define how things are done and when you can use certain operations and formulas. They explain what is and what is not allowed. They also define how a calculation will be performed.

Abstraction is the ability for one thing to represent something else. For example, imagine you live in a big city. A big city is an actual, real, thing. You can interact with it. You can drive through it. You can live, work, and play in it. Now imagine you need to drive to a part of the city you've never visited before. How would you chart your path? Many people would use a map, which is a visual representation (either on paper or on a computer) of the city. While a map is not a real city, it is a useful *abstraction*, or representation, of a real city. Mathematically, abstractions allow us to create very complex models and equations that represent how things look, work, or behave. Abstraction is extremely powerful and is one of the foundational mathematical tools used by both functions and equations. In fact, the differences between functions and equations are best explained by examining how each treats abstraction.

We will begin with an example that helps illustrate what abstraction is and how it works. Imagine for a moment that Helen and Hank are each holding five rocks. How many rocks do they have in total? Of course, we can easily show that together they have 10 rocks, because:

$$5\,rocks + 5\,rocks = 10\,rocks$$

This equation, which tells a story without using any real, physical rocks, illustrates an important characteristic about abstraction: Abstraction allows us to easily convey ideas without complexities that might be difficult or impossible to arrange in real life. In this case, we could determine that together they had 10 rocks without having to physically combine Hank's stack of "5 rocks" with Helen's stack of "5 rocks." In fact, the phrases "5 rocks" and "10 rocks" are nothing more than written text on the page you are reading. Words and numbers are not real rocks. Additionally, Helen and Hank might not even be real people. You may not know anyone named Helen or Hank; yet, we are able to construct a world, if only in your mind, where there are two people; one named Helen and one named Hank, who together have 10 rocks. Abstraction enables us to use something to represent something else. In this case, we used written numbers, like five, to represent quantities, and written words, like "rocks," to represent physical objects. The use of quantities to represent real things, concepts, and ideas is called numeric abstraction. **Numeric abstraction** forms the foundation of basic arithmetic and is easy to identify because it uses numbers to represent actual amounts or quantities.

Another important concept that is closely related to abstraction is called **types**. A type is a descriptor that tells us what something is or isn't. In our example, the type used in the story involving Helen and Hank is called "rocks." The mistreatment of types can often lead to incorrect concepts, conclusions, or mathematical errors. For example, if we had originally said that Helen has five rocks and Hank has five "sticks," then mathematical *rules* would prevent us from adding them together to conclude that they have 10 rocks in total, regardless of the fact that: $5 + 5 = 10$.

The incorrect treatment of types can lead to catastrophic mistakes. In 1999, the National Aeronautics and Space Administration (NASA) lost a spacecraft called the Mars Climate Orbiter. After a thorough failure analysis, NASA concluded that the spacecraft's disintegration was caused by incorrect mathematics resulting from a mismatch between types. Specifically, the thrust system used values that were in English units called pounds, while the navigation system used values that were in metric units called Newtons. So when the two systems talked with one another, they didn't take into account their respective types and necessary type conversions. In this case, the correct conversion is:

1 pound = 4.4482 Newtons

This mistake resulted in the lost of a $125 million spacecraft and adversely affected the $327 million orbiter and lander mission. Although mistakes can be made by the most disciplined and well–trained experts, types are highly useful and are treated with a high degree of rigor in mathematics, computer science, and engineering.

Returning to our example with Helen and Hank: Numeric abstraction allows us to use arithmetic statements "to abstract" things like the *quantity* of rocks when we know their specific amounts. We cannot write an arithmetic equation when we do not know the actual quantities involved. Notice that if Helen and Hank were going rock collecting *next* Saturday, we would not be able to use numeric abstraction. This shortcoming is overcome by introducing a new layer of abstraction, called **variable abstraction.**

Variable abstraction uses **variables** (eg, alphabetic letters) to represent amounts. Variables allow us to write useful math statements *now*, knowing that specific amounts will be filled in *later*. In effect, variables are placeholders.

Mathematically, the relationship between variables and numbers is very similar to the relationship between nouns and pronouns in the English language. Consider the following story:

> *She got on the bus. She sat next to me. She smiled. She saw that her stop was next and pressed the stop request button. She smiled again and stepped off the bus.*

While this story communicates the activities of a woman and what she did while on the bus, it is missing something. We are left with the unanswered question: Who is the woman? So, this story is not complete until we know exactly which woman we are talking about. We need to replace the pronoun with a noun. For example, if we knew that "she" (pronoun) was "your friend's mother" (noun), we would know exactly whom this story was about.

Mathematically, variables behave the same way: they serve as mathematical pronouns. For example, we can say that t is the total number of rocks Helen and Hank collect, x is the quantity of rocks in Helen's possession, and y is the quantity of rocks in Hank's possession. This allows us to tell a mathematical story that says

$$t = x + y$$

which is textually stated as *the total quantity of rocks is the sum of Helen's rocks added to Hank's rocks.* At the time the equation is written, we might not know their exact values. However, we can perform some interesting things with this equation, such as create a second equation that will determine the price that Hector will pay for their rocks. Imagine that Hector has agreed to buy their rocks for $5 per rock, which is represented by the equation:

$$p = 5t$$

Although we have two equations, we will not be able to determine how much Hector will ultimately pay Helen and Hank until *after* they have collected the rocks on Saturday. If on Saturday Helen and Hank collect five rocks each, Hector could use these equations to determine that he would need to pay a total of $50. In other words, the variables in the equations represent mathematical pronouns and the story is only complete once we know the actual values they hold. Variable abstraction is extremely powerful and is used to solve more complex problems than equations that use numeric abstraction alone.

Another example, one that we will revisit throughout the remainder of this chapter, is the algebraic equation:

$$y = x^2$$

which is written in the form

$$variable = expression$$

Algebraic equations, or **algebraic assignment statements**, are statements consisting of an expression that is assigned to a variable. This variable, when used with functions, is sometimes called an **instance variable**.

Because variables are treated differently by functions than by equations, we must explicitly show how each works with variables. To aid in our explanation, we are going to ask you to keep track of them by writing each variable on paper. Imagine you have several pieces of paper in front of you. One page is 8.5" by 11", called a letter–sized page. You also have several 3" by 5" index cards. On the top–left corner of the letter–sized page, write the words "global namespace." This letter–sized page represents the *global namespace*, which is a concept that we will develop shortly. For now, what you must know is that any time you use a variable as part of an equation, you must explicitly write it on the global namespace so that you can keep track of it.

You can write a variable only once on any given piece of paper. If the variable is already written, the meaning of that variable in each namespace is the union of its previous meaning when written on that namespace combined with how you are using it now. To help illustrate this point, imagine that you've hired a general contractor named Harry to work on your house. You later find out that Harry is also an electrician and ask him to do the electrical work. As fate would have it, you later learn that Harry is also a plumber and ask him to update your bathroom plumbing. Harry is the sum of all of the skills he possesses, even though you may have learned about his skills at different points in time.

The **global namespace** is the repository, or storage place, of all the variables that are used by an algebraic equation or collection of related algebraic equations. The act of using a variable in an equation automatically puts it into the global namespace. Figure 5–1 illustrates how the variables are placed on the global namespace and associated with the equation $y = x - vt$.

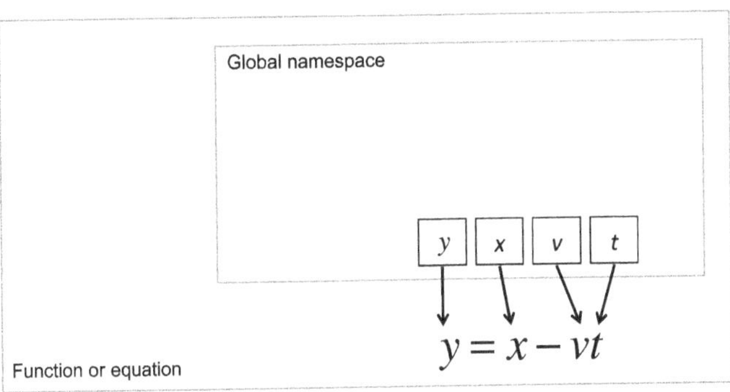

Figure 5–1 Variables on the global namespace associated with an algebraic equation.

Interestingly, namespaces are not generally covered in most mathematics courses. Because all variables used in algebraic equations are placed on the global namespace, algebra can be used quite effectively without understanding or defining what a namespace is and how it works. In fact, the introduction of namespaces would be superfluous. However, namespaces are extremely important and cannot be ignored when working with functions.

Functions are an extremely useful mathematical tool. Interestingly, most mathematical treatments of functions limit their discussion to the mapping of one set of information, called a *domain*, onto another set of information, called a *range*. While this treatment is useful, it does not sufficiently explain functions to the point where one can appreciate their potential and power. Fortunately, functions are thoroughly developed in computer science. A large amount of our understanding of why functions behave the way they do comes from the need to be specific when *teaching* a computer how to perform mathematical computations.

In Modern Mechanics, a **function definition** is an expression in the form:

$$function\_name(parameter\_list) = \{function\_body\}$$

An example function definition is:

$$f(x) = \{x^2\}$$

Structurally, a function definition has several parts: the **function name** (eg, f); the **function body** (eg, x^2, which is inside the brackets $\{\}$); and the **function signature**, which is the function name and any symbols or variables used as parameters by the function (eg, $f(x)$). Variables used in the function signature are called **parameters**, **function variables**, or **local variables**. The function body is alternatively called the **function equation** and the bracket notation is optional. While brackets are not used heavily in this chapter, this bracket notation is used throughout the text of this book to aid in distinguishing the function body in a function's definition from an algebraic expression. Since functions must always be invoked before they can be used, this notation helps to prevent us from forgetting this requirement.

Function invocation is the act of replacing a function's parameters with numeric values, variables, or expressions. Functions must be used in two steps. First, they must be defined. Then, they must be invoked before they can be used. Invocation is also called instantiation, and both terms are used interchangeably. In Modern Mechanics, a function invocation is often expressed in the form:

$$instance\_variable = function\_name(argument\_list)$$

The content between the parentheses is the **argument list**, also called **arguments**. The instance variable is simply a variable to which the instantiated expression is assigned. Arguments are numbers or variables in the global namespace that replace a function's parameters during invocation. Arguments can also be expressions made using numbers and variables on the global namespace.

Functions, like algebraic statements, are extremely powerful because of their use of *variable abstraction*. Functions can do everything algebraic equations can do. In addition, functions are able to solve more complex math problems than algebraic equations because of an additional layer of abstraction, called **namespace abstraction**.

Namespace abstraction is the ability of a function to create and use its own variables, enabling us to perform mathematical operations on numbers, other variables, or expressions, *without knowing what they are in advance*. For example, we could develop a function that will find a square, without knowing in advance the value, variable, or expression to be squared.

Conceptually, namespace abstraction is similar to the use of an additional descriptor to identify people. Imagine once again that you are building a house. To help you with the construction you have hired a general contractor, an electrician, and a plumber, where each person you've hired is named Harry. How would you distinguish between the three? You could refer to each by his first and last name. For example, you might refer to them as Harry Adams, Harry Brown, and Harry Carson, respectively. Alternatively, you might distinguish between them by appending their profession to their name. For example, you might refer to them as Harry the general contractor, Harry the electrician, and Harry the plumber. Regardless of the naming scheme, you are able to keep them separate and distinct from one another by adding a descriptor. In fact, if you fail to add this descriptor, you may return to your house to find Harry the electrician working on your plumbing based on your request: "Harry, can you fix the bathroom faucet?"

Mathematically, variables and symbols do not have last names or professions. Instead, we keep them straight and separate from one another by assigning them to specific namespaces. Namespace abstraction is, arguably, harder for some to initially understand, because variables with the same name can be used to represent different things. In order to facilitate our explanation of namespaces, we are first going to describe functions using **symbolic abstraction**.

As mentioned previously, functions, unlike an algebraic equation, are not ready to be used when first written. To use a function, two separate steps must be performed: function definition and function invocation. Let's examine the function definition, which is the first of the two steps. An example of a function definition, using symbolic abstraction, is:

$$f(\square) = \square^2$$

Unlike an equation, a *function definition* cannot be used as part of an equation or expression. Notice how the function uses a symbol, the white square, instead of a variable. The use of symbols in an equation instead of variables or numbers is called symbolic abstraction. To use this function, the white square must first be replaced with numbers or variables. *The only way to replace this symbol is through function invocation.*

A function invocation tells us how to replace the symbol. During invocation, the numbers or letters contained within the parentheses replace the symbols that were previously used in the function's definition. An example of a *function invocation* is:

$$y = f(5)$$

which replaces the white square with the number five to produce the **instantiated equation**:

$$y = 5^2$$

An instantiated equation is the equation produced after invoking a function. As a convenience to show that the instantiated equation originally began as a function, we use the ≡ operator instead of the = operator throughout this book.

Notice that the choice of the symbol we used in the function definition does not matter, since *it must be replaced during invocation*. In other words, a function defined as:

$$f(\lozenge) = \lozenge^2$$

that uses a white diamond instead of a white square will produce the exact same instantiated equations when invoked the same way.

Functions are considered equivalent if they have the same name and if, for every possible invocation, they produce the same instantiated equations. The symbols used in the definition are immaterial. Technically, these symbols cannot be assigned a value using an assignment statement. However, some historical texts do not adhere to modern invocation convention and incorrectly use an assignment operator to show the replacement.

Mathematically, functions can use all three types of abstractions: symbolic, algebraic, and numeric. This makes functions extremely powerful, the price being that they require the two steps of definition and invocation. Algebraic equations can use two types of abstractions: algebraic and numeric. Basic arithmetic can use numeric abstraction alone. While these three types of abstractions make sense, there is one significant complication that must be addressed: Variables are not limited to letters, and symbols are not limited to shapes.

Both variable and symbolic abstraction can use variables and symbols. Because algebraic and symbolic abstraction can use the same letters and symbols, we need a different technique to ensure that the behaviors we have just described work properly. This problem is overcome by using namespace abstraction instead of symbolic abstraction.

To understand how namespaces work, we have to re–examine how functions work. When a function is defined, it automatically creates a new namespace called a **local** *or* **function namespace**. (Technically, namespaces are created when a function is invoked. However, for the purposes of this discussion, assume that they are created when the function is defined.) Concurrent with the creation of a namespace, the symbols or variables that appear between the parentheses in the function definition create new variables in that function's namespace. Each function has its own unique and separate namespace. A function definition creates a new namespace and each variable contained within the parentheses is added to that namespace, effectively giving variables last names.

In a function definition, *anything* that is contained between the parentheses can conceptually be thought of as a symbol that must always be replaced during a function's invocation. In other words, we could use variables instead of symbols. In this case, think of x and y as symbols, not as variables. Mathematically, the function definitions $f(x) = x^2$, $f(y) = y^2$, $f(\square) = \square^2$, and $f(\Diamond) = \Diamond^2$ are equivalent to each other. The x, y, $\square$ or $\Diamond$ are placeholders in a function definition. These placeholders will be replaced with expressions during invocation.

When the functions are invoked as $y = f(a)$ and $y = f(b)$, they replace the symbols that appeared in the function definition. Since this replacement is a requirement, when any of the functions just mentioned are invoked as $y = f(a)$ they will produce the same instantiated equation, $y = a^2$. Similarly, when any of the functions just mentioned are invoked as $y = f(b)$, they will produce the same instantiated equation $y = b^2$. The replacement of the symbol used in a function's definition with a number, variable, or expression during invocation is what makes functions more powerful than algebraic equations. We will explain the mechanism behind this replacement shortly. However, what is important to recognize is that because this replacement is a requirement, we can only determine what global namespace variables the function actually uses once it has been invoked.

We now introduce two new terms: **local variables** and **global variables**. A *global variable* is a variable that is defined on the global namespace. Algebraic equations can only use variables that are defined on the global namespace. A *local variable* is a variable that exists within a function namespace and can only be used by that specific function. However, a function can also use variables from the global namespace.

As an example, consider the two statements:

$$x = 5$$
$$f(x) = x^2$$

The first statement is the equation $x = 5$, which automatically places a x variable on the global namespace (eg, the letter–sized page). The second statement, $f(x) = x^2$, is a function definition that automatically creates a new namespace and places an x variable on the newly created function namespace. To illustrate this step, imagine writing the "$f(x)$ namespace" on the top–left corner of one of the 3" by 5" cards to represent that function's namespace. Place that card on top of the global namespace (eg, place it on the letter–sized page). Second, imagine writing on the function namespace (the 3" by 5" index card) any variables that appear between the parentheses following the function name. In this case, you would write the x variable on the function namespace. As shown in Figure 5–2, *two x variables exist in two different namespaces and are separate and distinct from each other.*

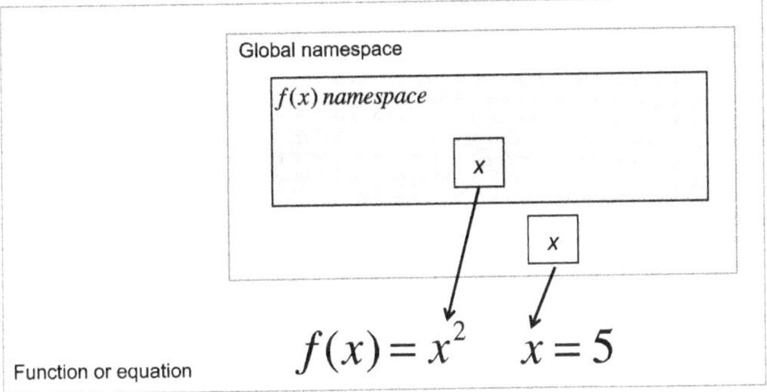

Figure 5–2 Function definitions create new namespaces and place any function's parameters into that namespace. Local parameters are called local variables. Variables with the same name can exist in multiple namespaces. When a variable has the same name in the global and function namespaces, they are different and distinct variables. Such variables are called overloaded.

As mentioned earlier, a function can use variables from its namespace as well as from the global namespace. For example, $f(x) = x - vt$ is a function definition that uses three variables. Since x is contained within the parentheses, we place an x variable in the $f(x)$ namespace. Because the other two variables, v and t, are not contained within the parentheses, they are placed on the global namespace. In the function $f(x) = x - vt$, the x variable refers to the variable that is contained on the $f(x)$ namespace, not the x variable placed on the global namespace by the $x = 5$ statement. Generally, a function cannot use a global variable that has the same name as a function's local variable. Figure 5–3 illustrates the relationship between global and local variables.

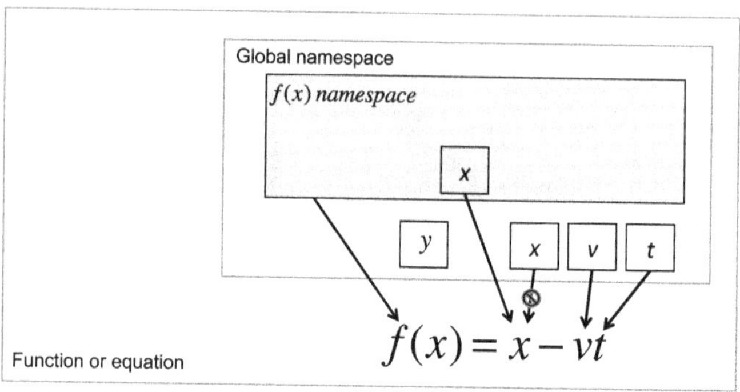

Figure 5–3 Functions can use variables from the global and function namespaces. When a function defines a variable that has a name in both namespaces, the variable in the function namespace takes precedence.

Non–nested functions, as used throughout this book, cannot use variables on another function's namespace, and algebraic equations can only use variables defined on the global namespace.

Variable abstraction and namespace abstraction uses the same set of symbols and letters to perform their abstraction. Because each namespace represents a unique and independent variable space, variables of the same name can exist in multiple namespaces. As a result, we can run into cases where the meaning is not clear and incorrect use is likely to occur. Recall the imagined situation of hiring three men named Harry to work on your house. As shared earlier, if you simply shout an instruction to Harry to work on your plumbing, you might later find that Harry the electrician is working on your plumbing. These conflicts are due to overloading, which occurs when multiple letters or symbols are used with different meanings. Mathematically, such variables are called **overloaded variables**.

Equations and functions that use overloaded variables can produce some of the most difficult–to–detect problems in all of mathematics. For example, consider only the x variable in the following two statements:

$$x = ct + vt$$
$$f(x) = x - vt$$

The statement $x = ct + vt$ places an x variable on the global namespace, while the function definition places an x on the local $f(x)$ function namespace. Notice that the x variable is considered **overloaded** because it exists in both the global and local namespaces.

A common mistake is to substitute x within the function definition with the expression $ct + vt$ from the algebraic equation $x = ct + vt$, resulting in $f = ct$ or $f(x) = ct$. This is incorrect and unfortunately an extremely common mathematical mistake. In fact, not only is this a common mistake, without an understanding of namespaces, it is nearly impossible to detect. The reason it is difficult to detect is because if it were an algebraic statement, it would be perfectly valid. Notice that:

$$x = ct + vt$$
$$y = x - vt$$

is a valid statement that will properly produce the equation $y = ct$ following a substitution of x with $ct + vt$. Remember that *the only way to place values into a function's local parameters is through invocation.* You cannot substitute values into a function's parameters because the variables are from different namespaces.

To avoid this overloading problem, variables can be written using their full names. The full name for a variable in the global namespace is formed by adding two colons (eg, "::") to the front of the variable. The full name for a variable in a function namespace is formed by adding the function name and two colons (eg, "f ::") to the front of the variable. This notation effectively distinguishes the global $::x$ variable in the global namespace from the local $f::x$ in the $f(x)$ namespace.

Demonstrating the use of formal notation by using the x variable only, both of the algebraic statements and the function definition would be written as:

$$::x = ct + vt$$
$$y = ::x - vt$$
$$f(f::x) = f::x - vt$$

This notation, while useful to identify which x variable is associated with which namespace, is cumbersome. While useful, especially in computer science, such notation can be hard to read. In practice, it is easier to simply refer to one variable as the "*local x*" and the other as the "*global x.*" Therefore, we can write it in a somewhat easier to read alternative form as:

$$global\_x = ct + vt$$
$$y = global\_x - vt$$
$$f(local\_x) = local\_x - vt$$

In effect, we are able to keep our variables distinct and separate from one another in a similar manner, as we were able to distinguish between our three contractors named Harry by adding their last name or profession.

Mistreatment of overloaded variables can also lead to subtle and extremely difficult–to–detect math mistakes. Such mistakes are nearly impossible to detect without formal namespace notation labeling. Problems associated with overloaded variables are some of the most difficult problems a computer scientist faces. In some cases, a computer scientist might find identifying the mistake so difficult that it is easier to rewrite the equations, or algorithm, from scratch using different variables.

Function names can also be overloaded. For example, the functions $f(x) = x - vt$ and $f(x,t) = x - vt$ overload the function f. It is not clear to which function we refer when we use the function name, f, alone. Notice that the first definition defines a function with one local parameter, x, which will be replaced during invocation. The second defines a function with two local parameters, x and t, that will be replaced during invocation. Fortunately, because both function definitions use a different number of parameters, the correct function would easily be identified upon invocation. For example, we would have no trouble knowing which function to invoke if the invocation statements were $y = f(5)$ and $y = f(5,1)$. In this case, we determine which function to invoke by matching the number of arguments in the function's invocation to the number of parameters used in the function's definition. However, function overloading remains a problem when the overloaded functions have the same number of parameters. For example, $f(x) = x^2$ and $f(x) = x - vt$ are two functions with one parameter. Upon invocation with the statement $y = f(5)$, we would not know which function to use. Generally, two functions with the same name and number of parameters cannot be used within the same system of equations.

Functions and equations are mathematically different because of how they work with variables and namespaces. To illustrate this difference, consider the algebraic equations $y = x^2$, $y = m^2$, and $y = s^2$, which are mathematically *different*. They use different variables in the global namespace. As a result, they are not equivalent and cannot be used interchangeably.

Now consider the three function definitions $f(x) = x^2$, $f(m) = m^2$, and $f(s) = s^2$. Because functions use namespace abstraction, these three functions are mathematically equivalent, even though they use different variables in their definitions. Due to a function's use of namespaces, function equivalence is based on what happens during invocation, not how their definitions are written. Following function invocation with the same arguments, each will produce the same instantiated equations. Function equivalence is determined by what functions do when invoked with identical arguments, not solely on what they look like when defined. Because the three functions above will always produce the same instantiated equations, they are equivalent and can be use interchangeably.

Equations are not equivalent to other equations if they use different variables. Functions are equivalent if they use have the same name, take the same number of arguments, and produce the same instantiated equation for every possible invocation.

Figure 5–4 illustrates the relationship between six functions and equations. Notice that $y = x - vt$ and $f() = x - vt$ are equivalent, and so are $y = a - vt$ and $f() = a - vt$. While equivalent, the function definition must still be invoked in order to transform it into an instantiated equation. Equations and functions are considered equivalent if they use the same variables in the global namespace and the function takes no arguments.

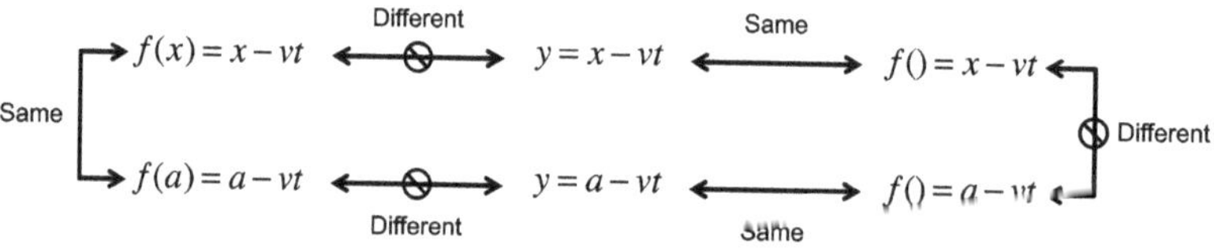

Figure 5–4 Relationship between functions and equations that reflect how each uses variables on the global and function namespaces.

Summary

When used properly, functions are one of the most powerful mathematical tools available. Mathematics gains its power in large part by how it leverages abstraction. This chapter introduced the important concepts associated with functions, including namespaces, local and global variables, function definition versus invocation, function signatures, function parameters and arguments, and variable and function overloading. Function behavior, which is more easily described using these terms, is part of the computer science body of knowledge, but also applies to functions in engineering and mathematics.

Arithmetic, algebraic equations, and functions each introduce new capabilities and power based on how they use abstraction. Algebraic equations and functions both use variable abstraction and namespaces. Algebraic equations, which only use the global namespace, are ready to use when they are written. Functions, which can use multiple namespaces, are not ready to be used when first written. Functions require two steps: definition and invocation. A function cannot be used until it has been invoked. Often, equations and functions are generally treated synonymously. However, because of the nuances associated with namespaces, problems can arise, especially those associated with overloaded variables. This is a subtlety that is often overlooked by people less familiar with functions.

Because functions create their own namespaces, *a variable t with the same name might exist in different namespaces, with each one being uniquely different from variables of the same name in other functions and equations.* Using the same variable in a system of equations, where they are technically different variables, can be quite confusing. Problems with overloaded variables are some of the most difficult to resolve in computer science and mathematics.

In many ways, this introduction to functions and namespaces only scratched the surface of their potential power. We did not discuss all of the possible nuances and exceptions that can occur, and only went to a depth required to understand the material presented in this book. For readers with a deeper interest in functions, please review *Compilers: Principles, Techniques, and Tools* by A. Aho and M. Lam. In addition, many books about object–oriented programming will contain a discussion of functions. While material associated with functions and namespaces will be familiar to many computer scientists, they have originally learned this material using different terminology. For example, computer scientists may be more familiar with terms like "the heap," which is similar to the global namespace, and "the stack," which is similar to a function's namespace.

Chapter 6 Relativity and its Critical Mistakes Explained

At the turn of the 20th century, physics had arrived at a crisis. Classical mechanics could not explain experiments involving electromagnetic force and optics. The introduction of relativity theory filled a gap in the scientific landscape. Relativity was as ingenious as it was revolutionary and owes a large part of its success to the fact that it explained certain experiments better than any of its contemporaries, including classical mechanics. That statement bears repeating: Relativity could mathematically explain what scientists were observing, which is something classical mechanics was unable to do.

Since the publication of *Zur Elektrodynamik bewegter Körper (On the Electrodynamics of Moving Bodies)* in 1905, relativity theory has become the cornerstone of modern physics and is one of the most widely taught theories of the 20th and early 21st centuries. Like the ancient Greek model of the solar system, relativity theory has served mankind well for a long time; and like the ancient Greek model of an Earth–centered universe, relativity theory is one of the most well–recognized theories that happens to be completely wrong.

Relativity has survived as a leading theory because, with the Type I error in Einstein's proof overlooked, the derivation, concepts, and theory were thought to be correct. We have already shown in Chapter 1 that Einstein's spherical wave proof failed. This failure means that there is something wrong with relativity theory. We must now show exactly what is wrong with the theory, despite it appearing to work well. We must also answer the question: *If Einstein's work is wrong, why does relativity work so well and provide better answers than its classical mechanics counterpart?*

The short answer to this question is simple: *Einstein's equations do not explain what Einstein thought they explained.* With an understanding of what his equations really explain, one will be better able to see why his equations perform better than the classical mechanics–based equations and why his equations perform worse than the Modern Mechanics equations. In short, relativity theory is a combination of Einstein's postulates, key concepts, unique math mistakes, and incorrect explanations that produced equations that often give acceptable answers.

In this chapter, we will reveal where Einstein makes several critical conceptual and mathematical mistakes, and why his spherical wave proof was so critical to the establishment of his theory. Because Einstein's proof has failed, we are free to explain relativity theory on a level playing field, unencumbered by the limits, interpretations, and terminology of Einstein's work. Instead, we will examine Einstein's assumptions, derivation, and key implications using the key concepts and terminology of Modern Mechanics. This will aid in our explanation and allow us to reveal aspects of Einstein's work that have been either misunderstood or ignored.

Types of Math Errors

Math mistakes can take several forms and generally fall into two categories: syntax errors or semantic errors. Syntax errors are also called rule violations. Syntax errors can be obvious and easy to find, or they can be non obvious and difficult to find. For example, an obvious rule violation is the statement:

$$2+2=5$$

which is clearly incorrect because the sum of two plus two is four, not five. Other syntax errors are harder to find because, although they are incorrect, they produce the right answer. For example, one might simplify $\frac{16}{64}$ by canceling the sixes, to produce $\frac{1}{4}$.

While $\frac{1}{4}$ is the correct answer, the method used to arrive at the answer is incorrect and generally will not produce the right answer. In this case, the correct answer is to remove the common factors from the expression. When the numerator is divided by 16, which is the largest common factor, it simplifies to 1 and when the denominator is divided by 16 it simplifies to 4, correctly producing the answer, $\frac{1}{4}$.

Semantic errors are mistakes of context or meaning. These errors are generally hard to detect because the individual syntax of each statement may be valid, yet the series of statements in its entirety is incorrect. This often occurs because an earlier statement makes one of the subsequent statements invalid. An example of a semantic error is the proof establishing that 1 = 0, which is:

1. Begin with: $x=y$
2. Multiply by x: $x^2=xy$

3. Subtract y^2: $x^2 - y^2 = xy - y^2$

4. Factor: $(x + y)(x - y) = y(x - y)$

5. Divide by $(x - y)$: $x + y = y$

6. Substitute for x since $x = y$: $y + y = y$

7. Simplify: $2y = y$

8. Divide by y: $2 = 1$

9. Subtract one from each side: $1 = 0$

Each step, when evaluated on its own, is a proper math statement. In other words, from a syntax perspective, this proof appears to be correct. However, because each statement in this proof must always be true, we can find where a semantic error occurs. Step 1 says $x = y$ is true, which means that $x - y = 0$. Therefore Step 5, which divides both sides by $x - y$, is a rule violation, because division by zero is not allowed. Hence the proof actually failed. Although semantic errors are harder to detect, their appearance in any work is as much of a mistake as the more obvious syntax errors.

The Newtonian Transformations

Chapters 2 through 4 developed the conceptual and mathematical framework for the Modern Mechanics model. In Modern Mechanics, a two–system model uses the geometric transformations of translation, rotation, scaling, and reflection to explain the motion of the inner system. Modern Mechanics and classical mechanics share similar foundations. However, as discussed in chapters 2 and 3, classical mechanics is a two–

system model that cannot explain oscillating motion in moving systems. Oscillating behavior is explained in Modern Mechanics using a third system, called the oscillating system. The oscillating system moves back and forth between two points on the inner system. The relationship between the oscillating, inner, and outer systems can be nested or non–nested, depending on the placement of the oscillating system. In a nested system relationship, the oscillating system is placed on and moves with respect to the inner system. In a non–nested relationship, the oscillating system is placed on and moves with respect to the outer system.

Modern Mechanics works with many different types of waves and mediums. In contrast, relativity theory was developed for one medium: the electromagnetic force. The propagation of an electromagnetic force in a vacuum occurs at the specific velocity, c, or 299,792,458m/s. Because Einstein's work was developed specifically for electromagnetic force, we will use his variable, c, in this chapter instead of the general variable for a propagating wave, w.

Modern Mechanics uses the terms inner system, outer system, and oscillating system. While Einstein does not use these exact terms, he uses other terms with similar meanings. An outer system is synonymous with Einstein's stationary system and an inner system is synonymous with his moving system. In fact, in this chapter these terms will be used interchangeably with their counterparts, ideally to promote overall comprehension of the material.

The first case we will examine is where the inner system is in uniform translatory motion with respect to the outer system. Using one of Modern Mechanics' five rules, we place an inner system on an outer system. "Uniform translatory motion" simply means that the inner system moves according to the translation equation in one direction at a constant velocity. Since the inner system is in translatory motion, the translation transformation, or Newtonian equation, always applies.

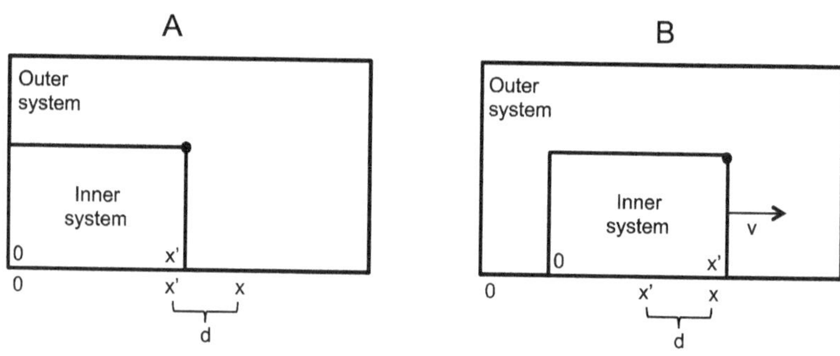

Figure 6–1 The inner (or moving) system moves in uniform translatory motion with respect to the outer (or stationary) system according to Equation 6.1. The dimensions of the inner system are x', y and z. The placement of x and x' are intentional.

As illustrated in Figure 6–1, the moving system starts at position x' and arrives at position x. Notice that the use of x and x' in Figure 6–1 differs from how they are used by both classical and Modern Mechanics. This variable use is not a mistake. In this case x' represents the starting position and x the ending position with respect to the outer system. The relabeling of the variables here is done to align with Einstein's 1905 paper, which is analyzed in the remainder of this chapter. The moving system moves with respect to the stationary system according to Equation 4.12, and Einstein reversed the use of the variables so that the translation transformation is written as:

$$x = x' + vt_v \qquad \text{Eq. 6.1}$$

where v is the velocity of the moving system and t_v is the amount of time that the moving system has been in motion. It is important to recognize that, because the moving system always moves with respect to the stationary system according to the translation equation, *we can always use this equation to determine the position of the moving system with respect to the stationary system.*

Generally, we use the Newtonian equations where time is increasing in value, beginning at $t_v = 0$; but this is not a requirement for using the equation. When you know where the system is at time t_v, you can use this equation to determine the moving system's historical position, when $t_v = 0$. In this case, we move the vt_v term to the other side of the equation and rewrite it as:

$$x' = x - vt_v \qquad \text{Eq. 6.2}$$

This is graphically illustrated in Figure 6–2 where the inner system is shown with respect to the outer system at time t_v and then we show where the inner system started when $t_v = 0$.

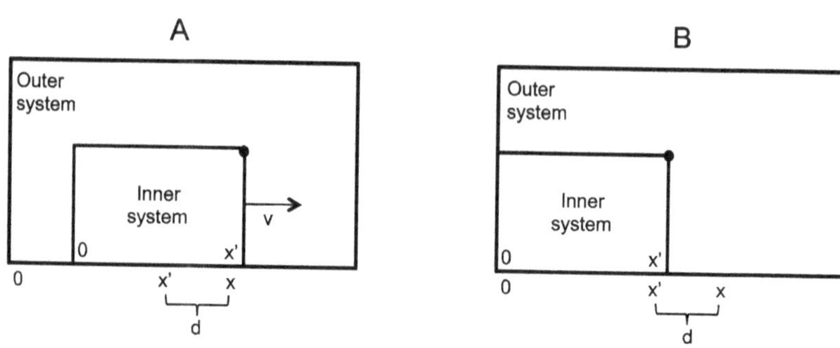

Figure 6–2 As illustrated in (A), the inner system at time t_v is at position x with respect to the outer system after it has moved to the right at velocity v. (B) illustrates the historical position of the inner system when time $t_v = 0$, when it was at position x' with respect to the outer system. Because the origins are coincident at $t_v = 0$, we can also determine the dimensions of the inner system.

This placement of the inner system's origin to be coincident with the outer system's origin to determine its dimensions is exactly what happened when Einstein says:

> *"If we put $x' = x - vt$, then it is clear that a point at rest in the system k [the moving system] has a definite, time–independent set of values x', y, z belonging to it."*

In other words, Einstein begins by assuming that the inner system has been in motion for time t. (From this point forward in this chapter, to align with Einstein's choice of variables, we use t, instead of t_v to represent the amount of time that the moving system has been in motion.) Einstein uses the translation transformation in reverse by taking the current position of the moving system at time t and finding its historical position when $t = 0$. By performing this translation, Einstein has found the position of an arbitrary point at position (x', y, z) with respect to the inner system. In addition, he has determined the dimensions of the moving system.

As illustrated in Figure 6–3, each of the variables in $x' = x - vt$ are in the global namespace.

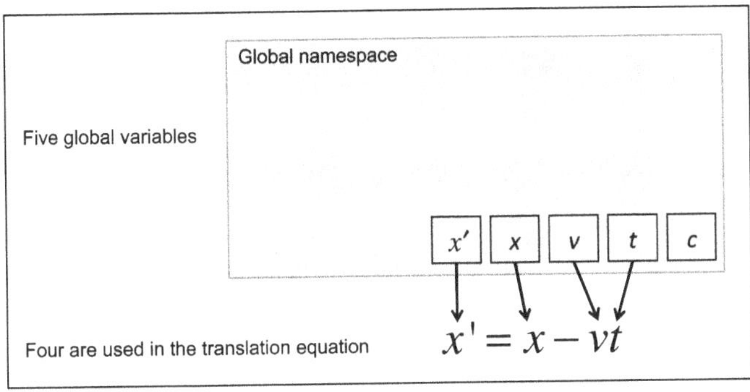

Figure 6–3 Namespace identification of the variables used in Einstein's translation equation.

In a non–nested relationship in Modern Mechanics, the inner system moves with respect to the outer system. Because Einstein used the translation equation to define the relationship between the moving and stationary systems, relativity is no different. One of the most overlooked characteristics of relativity theory is that Einstein states the moving system always moves with respect to the stationary system using the translation transformation, also known as the Newtonian equation. Specifically, Einstein overlooked the fact that he used $x' = x - vt$ as the translation transformation, which *always applies to the moving system.* This mistake leads to a cascade of conceptual and mathematical mistakes in the remainder of Einstein's work. It also presents a very important implication: Because the Newtonian equation always applies, Einstein's transformation equations cannot explain the position of the moving system. They must explain something else. We must now show what his equations actually explain and how Einstein's mistake affects the remainder of his derivation.

To aid in our discussion, assume that the inner system is a three–dimensional rectangular shape, called a cuboid, with the dimensions (x',y,z) on which we name the eight points representing its corners. The positions of the points with respect to the inner system do not change, regardless of the inner system's velocity. These points are:

1. Point P is at the origin, $(0,0,0)$.
2. Point Q is at $(x',0,0)$.
3. Point R is at $(0,y,0)$.
4. Point S is at $(0,0,z)$.
5. Point T is at (x',y,z). This is the point that Einstein references in his statement about the translation transformation.
6. Point U is at $(x',y,0)$.
7. Point V is at $(x',0,z)$.
8. Point W is at $(0,y,z)$.

The Oscillating System

In Modern Mechanics, back–and–forth motion is explained using a third system that oscillates between two points on the inner system. While Einstein does not explicitly use a third system, one is involved in his derivation. In Einstein's work, the *ray of light* plays the role of the third system. Einstein says:

> *"From the origin of system k, let a ray be emitted at the time τ_0 along the x–axis to x', and at the time τ_1 be reflected thence to the origin of the coordinates, arriving there at the time τ_2."*

This statement is synonymous with the back–and–forth motion of an oscillating system between the point P (the origin) and point Q of the inner system. The oscillating system, or ray of light, is emitted from the origin of the inner system and travels along the path of the forward segment on the x axis from point P to Q. When it arrives at point Q, it has traveled the forward intercept length. The amount of time required for it to travel from P to Q is defined as the forward intercept time, which Einstein calls τ_1. Upon arriving at point Q, the ray of light is reflected back and travels along the path of the reflected segment from Q to P. Upon reaching the origin of the inner system, the length of the return path is defined as the reflected intercept length. The time it took to travel from Q to P is the reflected intercept time. The total oscillation time, which Einstein denotes as τ_2, is the sum of the forward and reflected intercept times. Although relativity and Modern Mechanics use different terminology, Einstein uses the ray of light as a third, or oscillating, system. Because this ray can behave as part of a nested or non–nested relationship in Modern Mechanics, we must determine which type of relationship model Einstein uses in his work.

Einstein's theory is based on two assumptions, or "postulates." A postulate is the scientific term for an assumption that must always be true. The first of these postulates is called *the principle of the constancy of the velocity of light*, which states that:

> *"Every ray of light moves in that "rest" coordinate system with a fixed velocity c, independently of whether this ray of light is emitted by a body at rest or in motion."*

Einstein is clear that the ray of light moves with respect to the "rest," or stationary system. He also says that the velocity of the ray of light, with respect to the stationary system, is independent of the moving system's velocity. In other words, the ray of light moves with respect to the outer, or stationary, system at velocity c. According to Einstein's text, both the moving system and the ray of light move with respect to the stationary system using geometric transformations. The ray moves at velocity c and the moving system moves at velocity v. In other words, Einstein's *principle of the constancy of the velocity of light* postulate textually defines a non–nested relationship between the stationary system, the moving system, and the ray of light.

Einstein has a three–system model involving a stationary system, a moving system, and a ray of light. We can use the equations developed in Chapter 4 to find the forward and reflected intercept lengths and times. Because the distance between point P (the origin) and point Q of the inner system is x', the forward intercept time is:

$$\frac{x'}{c-v} \qquad \text{Eq. 6.3}$$

As long as the velocity of the moving system is less than that of the ray of light, the ray will be able to complete the journey from P to Q. If oscillations are required in Einstein's work, then this equation will establish an upper limit on the velocity of the moving system. We can also determine the reflected intercept time for the reflected ray traveling from Q to P as:

$$\frac{x'}{c+v} \qquad \text{Eq. 6.4}$$

The ray is able to complete the journey from Q to P regardless of the velocity of the moving system. The non–nested relationship between the stationary system, moving system, ray of light, and the corresponding forward and reflected intercept times are illustrated in Figure 6–4. These forward and reflected intercept times are important equations that will appear later in Einstein's derivation.

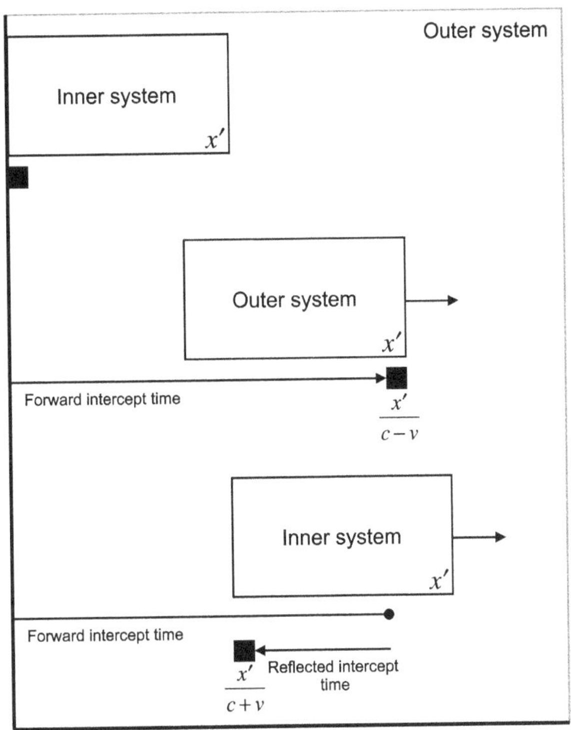

Figure 6–4 Relationship between the outer system (stationary or at rest system), the inner system (moving system), the oscillating system (ray of light), and the corresponding forward and reflected intercept times for the ray traveling along the x axis. The ray travels at velocity c and the inner system travels at velocity v.

While we have found the forward and reflected intercept times for the ray of light in Einstein's work, we can also produce the average intercept time. As discussed in Chapter 4, the average intercept time is found using the **addition mean equation**, which sums the forward and reflected intercept times and then divides that sum by two, producing:

$$\frac{cx'}{c^2 - v^2} \qquad \text{Eq. 6.5}$$

Alternatively, the average intercept time is found using the **subtraction mean equation**. To find the average intercept time using the subtraction mean equation, we must know the forward or the reflected intercept time, and the intercept time half–difference. We will find the intercept time half–difference in two ways. First, returning momentarily to the variable notation used in Chapter 4, the intercept time half–difference is:

$$t_H = \frac{xv_x + yv_y + zv_z}{w^2 - v_x^2 - v_y^2 - v_z^2} \qquad \text{Eq. 6.6}$$

To align this equation with Einstein's variables, we replace the general velocity of a wave, w, with the velocity of the ray of light, c, and we use x' instead of x, resulting in:

$$t_H = \frac{x'v_x + yv_y + zv_z}{c^2 - v_x^2 - v_y^2 - v_z^2} \qquad \text{Eq. 6.7}$$

As stated earlier, Einstein's derivation presumed *uniform translatory motion*, which means that he only considered the case where the moving system moves along the x axis in one direction at a constant velocity. We account for this constraint by setting the velocity components v_y and v_z to zero. While not a limitation in Modern Mechanics, this directional constraint is one that Einstein will later seek to correct with general relativity. Notice that when we consider only velocity along the x axis, the intercept time half–difference equation simplifies to:

$$t_H = \frac{x'v_x}{c^2 - v_x^2} \qquad \text{Eq. 6.8}$$

Because we only consider velocity of the moving system in one direction, we can replace v_x with v to represent the velocity of the moving system, resulting in the equation:

$$t_H = \frac{vx'}{c^2 - v^2} \qquad \text{Eq. 6.9}$$

With the intercept time half–difference known, the average intercept time is found by either adding it to the reflected intercept time or by subtracting it from the forward intercept time.

Alternatively, the intercept time half–difference is found using the forward and reflected intercept expressions used in Einstein's work. Returning momentarily to the notation developed in Chapter 4 for the subtraction mean equation, the intercept time half–difference, U_H, is found by subtracting the reflected intercept time from the forward intercept time and dividing the result by two. Using the forward and reflected intercept time expressions from equations 6.3 and 6.4, the intercept time half–difference is:

$$U_H = \frac{vx'}{c^2 - v^2}$$

With the intercept time half–difference known, the average intercept time, U_A, is found by subtracting the intercept time half–difference, U_H, from the forward intercept time, U_B, such that:

$$U_A = U_B - U_H$$

The subtraction mean equation will also produce the average intercept time when the intercept time half–difference is added to the reflected intercept time, U_S, such that:

$$U_A = U_S + U_H$$

Figure 6–5 illustrates the use of the subtraction mean equation to find the average intercept time using Einstein's variable notation.

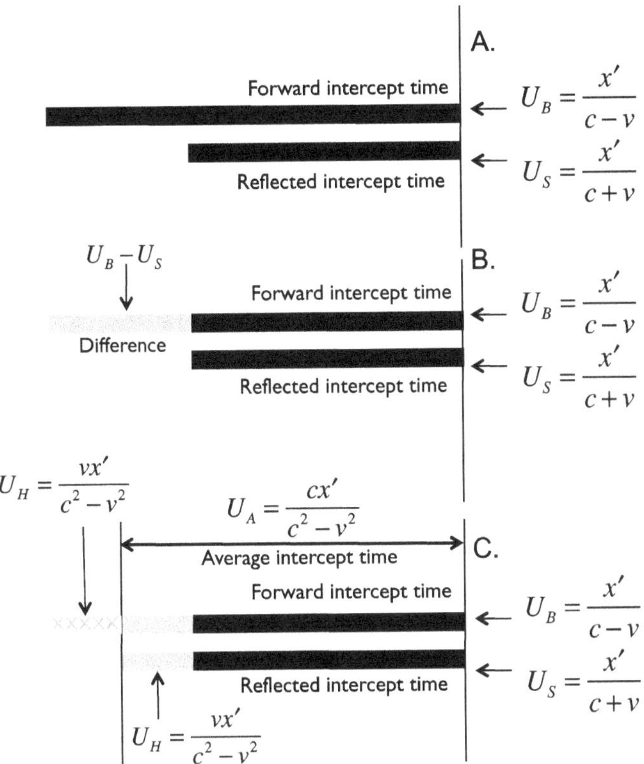

Figure 6–5 Illustration of the relationships between the forward intercept time, the reflected intercept time, the average intercept time, and the intercept time half–difference used in the subtraction mean equation. The average intercept time is found when the intercept time half–difference is subtracted from the forward intercept time, or when it is added to the reflected intercept time.

Notice that the intercept time half–difference, $\frac{vx'}{c^2-v^2}$, appears in the body of Einstein's Tau function, which we write formally as:

$$\tau(x',y,z,t) = \left\{ \alpha(t - \frac{vx'}{c^2-v^2}) \right\} \qquad \text{Fn. 6.1}$$

While we have shown that Einstein uses the intercept time half–difference in his Tau function, we must now examine Einstein's Tau function to determine its behavior and purpose. Specifically, we must determine whether Einstein's Tau function uses the intercept time half–difference as part of the subtraction mean equation to return the average intercept time.

Understanding Einstein's Tau Function

Einstein's τ, or Tau, function has taken on almost magical properties. Often described as "Einstein's time adjustment," it is one of the least–understood aspects of his 1905 derivation. Without a clear understanding of what the function does and what each of its parameters means, we cannot begin to understand its purpose or graphically illustrate its behavior; let alone appreciate or understand the power of the function and the role it plays in the remainder of Einstein's work. To understand the remainder of Einstein's derivation, we have to remove the mystery from this "magical" function and explain its purpose.

"I THINK YOU SHOULD BE MORE EXPLICIT HERE IN STEP TWO."

Einstein tells us in several different places that Tau is a function. He defines it as a function that takes four arguments by textually saying to "*first define* τ *as a function of* x', y, z, *and* t." This statement is then supported mathematically when Einstein invokes Tau during his derivation. Consider the first three invocations included in his statement:

$$\frac{1}{2}\left[\tau(0,0,0,t)+\tau\left(0,0,0,t+\frac{x'}{c-v}+\frac{x'}{c+v}\right)\right]=\tau\left(x',0,0,t+\frac{x'}{c-v}\right)$$

Eq. 6.10

where it is easy to show that each of Einstein's three explicit Tau invocations take four arguments.

Einstein uses his Tau function before he has defined the function body. Said differently, he has invoked Tau without knowing what the function does or how it does it. At this point in Einstein's derivation, Tau is an unknown function. While it sounds backward to invoke a function before you know what it does or how it works, it is not uncommon. In fact, it is such a common occurrence that disciplines like mathematics, computer science, and engineering provide tools, like reverse engineering and partial differential equations, to help us determine what an unknown function really does. Einstein used a mathematical tool called a partial differential equation (PDE) to define his Tau function. PDEs are used when you know how a function is invoked, but you don't know how it behaves. As a result of the PDE used to find Tau, Einstein says:

"Since τ is a linear function, $\tau = \alpha(t - \frac{vx'}{c^2 - v^2})$ "

Technically, the mathematical statement in Einstein's sentence is incorrect, because the result of a partial differential equation is a function, not an equation. By modern convention, Einstein's Tau statement is written as an equation that simply performs an assignment of an expression into the τ *variable*, not as a function definition. Without the preamble "*Since τ is a linear function,*" Einstein's Tau statement is easily confused as a variable assignment statement or equation. The Tau function should have been written as:

$$\tau(x',y,z,t) = \left\{\alpha(t - \frac{vx'}{c^2 - v^2})\right\} \qquad \text{Fn. 6.2}$$

which properly defines Tau as a function that takes four arguments. Like all functions, Tau consists of a function name, τ;

a function signature consisting of its name and parameters, $\tau(x',y,z,t)$; and a function body, $\left\{\alpha(t-\frac{vx'}{c^2-v^2})\right\}$.

As a function, Tau is part of the global namespace, but its variables are not. Variables defined in the function's signature are part of the function's namespace, not the global namespace. This is an extremely subtle and important characteristic that is often overlooked in analyses of Einstein's work. Figure 6–6 illustrates the namespaces once the Tau namespace and its local function variables are defined.

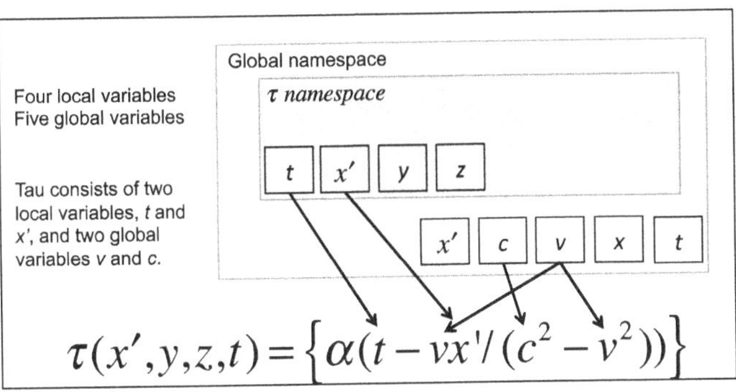

Figure 6–6 The Tau namespace, which contains its local variables, is used in conjunction with the global namespace. Notice that when overloaded variables exist (variables of the same name in the global and local namespace), the local variables take precedence within the function.

Although Einstein has defined the function body, he has still not explained what the function does. To determine its behavior, we have to use the second technique of reverse engineering. Einstein invokes Tau six times as part of his derivation, three times implicitly and three times explicitly. We must examine each invocation to determine Tau's purpose.

Prior to explicitly invoking Tau three times, Einstein said:

$$\frac{1}{2}(\tau_0 + \tau_2) = \tau_1$$

which defines three instance variables. We use this statement in conjunction with his invocations to assign each instance variable to its corresponding function invocation, such that:

$$\tau_0 = \tau(0,0,0,t),$$
$$\tau_2 = \tau(0,0,0,t + \frac{x'}{c-v} + \frac{x'}{c+v})$$

and

$$\tau_1 = \tau(x',0,0,t + \frac{x'}{c-v})$$

We can now examine each of Einstein's invocations to determine exactly how Tau works and explain each of its local variables. As illustrated in Figure 6–6, the Tau function body consists of both global and local variables. Because Tau is a function, it is critical to differentiate between the global variables that are used as *arguments* in each function invocation and the local variables that are used as *parameters* in the function definition. We will begin to reverse engineer Tau by exploring the meaning of the fourth parameter.

Einstein says:

> *"From the origin [of the moving system at time τ_0] the ray of light travels along the x axis to [point Q at] x' [of the moving system, arriving at time τ_1]."*

This textual statement is mathematically stated in Einstein's second and third function invocations. In these invocations, he uses the forward intercept time, $\frac{x'}{c-v}$, to represent the time required for the ray to travel from the origin to point Q. Einstein also says that when the ray of light reaches point Q, it is "*reflected thence to the origin.*" So, in his second invocation, he also uses the reflected intercept time, $\frac{x'}{c+v}$, which represents the amount of time required for the ray to travel from point Q to the origin. These invocations are extremely important for two reasons. First, they reveal the use of the forward and reflected intercept times in invoking Tau. Second, their use requires that the moving system be in motion according to the laws of the Newtonian transformation. This is confirmation that *the ray of light and the moving system both move with respect to the stationary system as part of a non–nested relationship according to the rules of geometric transformations.*

The function body, $\left\{\alpha(t-\frac{vx'}{c^2-v^2})\right\}$, consists of both local variables and global variables. Einstein later shows that α is 1, allowing us to safely ignore it in the remainder of our analysis. The existence of the intercept time half–difference:

$$\frac{vx'}{c^2-v^2}$$

in the function body suggests that the Tau function uses the subtraction mean equation to produce its result.

As discussed in Chapter 5, functions can create a unique situation – overloaded variables – that never exists when solely using equations. Overloaded variables are ones that have the same name, but actually have different uses and represent different variables. Imagine for a moment two men, both named Harry. If you only refer to them by their first names, you might confuse them for one another. To avoid confusion, you may need to refer to them by their full, or formal, names. When variables are overloaded, local variables must be distinguished from global variables of the same name. The role of a function's local variables can, and often does, differ from the role of a global variable with the same name.

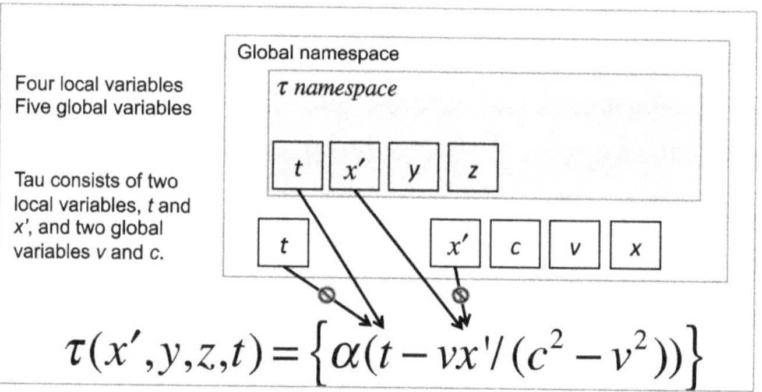

Figure 6–7 Global and local variables. Notice the distinction between the local and global t variables, and the local and global x' variables. When treated algebraically, one will not detect that t is overloaded, leading to the function's incorrect use as an equation.

As illustrated in Figure 6–7, Einstein's Tau function consists of both global and local variables, meaning that it uses variables from both the global and function namespaces. Two of the variables, t and x' are overloaded, meaning they exist in the global and the function namespaces. Inside the function signature and function body, the t and x' variables are local variables. Values are "inserted" into local variables during a function's invocation. This is an important characteristic that distinguishes functions from equations. Care must be taken when performing substitutions in functions comprised of overloaded variables.

Because Tau is a function, the t variable that is used <u>outside</u> the function signature and function body is different from the one used inside the function signature and function body. Outside of the function, t is the time that the inner system has been in motion. Because the t inside the function does not need to hold this meaning, we must determine what the local t variable represents. If it represents the forward intercept time, then, through the definition of the subtraction mean equation, we will be able to conclude that the Tau function returns the average intercept time.

The global t variable appears in each of Einstein's explicit Tau invocations. As a global variable that is used as a function argument, it has the same value and meaning during the function's invocation as when used in the Newtonian equation, where it represents the time that the moving system has been in motion. However, within the function definition where it is a function parameter, it is a local variable and the global t variable plays no role in his resulting Tau function. Because the global t variable does not play a role in the Tau function body, we will simplify our analysis by setting $t = 0$ in the function invocation. This simplification will enable us to better examine the role and behaviors of the local variable t within the function body. Once the analysis is complete, we will examine the significance of this simplification, as well as explore what happens when its value is not zero.

The first invocation we will examine is Einstein's explicit invocation:

$$\tau(x',0,0,\frac{x'}{c-v})$$

which contains the forward intercept time as its fourth argument. The primary activity that occurs during function invocation is the insertion of arguments, or more specifically, their expressions, into the local variables. In this invocation, the forward intercept time, $\frac{x'}{c-v}$, is inserted into the local variable t.

This use of namespaces is an important nuance that distinguishes functions from equations. The insertion of the arguments into the local variables means that the invocation of the Tau function as:

$$\tau(x',0,0,\frac{x'}{c-v})$$

produces the instantiated expression

$$\equiv \frac{x'}{c-v} - \frac{vx'}{c^2-v^2}$$

The instantiated expression subtracts the intercept time half–difference from the forward intercept time. We have already shown that the subtraction of the intercept time half–difference from the forward intercept time produces the average intercept time. This equation simplifies to the x axis average intercept time, or:

$$\equiv \frac{cx'}{c^2-v^2}$$

In other words, this analysis of Einstein's Tau invocation reveals that Tau returns the average intercept time for a ray oscillating between the origin and point Q at $(x',0,0)$ of the moving system.

The second Tau invocation we will explore is the same as the first, with one difference: It is invoked implicitly rather than explicitly, partially masking the function's invocation. The implicit invocation occurs when Einstein says:

$$\xi = c\tau$$

Notice that in this statement and its corresponding invocation, the τ function is incorrectly shown as a variable. Einstein should have written this statement as $\xi = c\tau_1$, using the Tau instance variable he previously defined. He could have also invoked the function by writing it informally as $\xi = c\tau()$ or explicitly as $\xi = c\tau(x', 0, 0, \frac{x'}{c-v})$.

Because ξ is the product of the average intercept time for the ray of light multiplied by the velocity of the ray of light, it must represent the average intercept length for the ray oscillating between the origin and point Q. Einstein performs his informal invocation when he says to *"insert this value* $\left[\frac{x'}{c-v}\right]$ *of t in the equation for* ξ," to produce the average intercept length for a ray traveling along the x axis, or:

$$\xi = \frac{c^2 x'}{c^2 - v^2} \qquad \text{Eq. 6.11}$$

which simplifies to:

$$\xi = \frac{x'}{1 - \frac{v^2}{c^2}} \qquad \text{Eq. 6.12}$$

Unfortunately, Einstein shows this time "insertion" using an assignment operator, which is inappropriate for a function variable, which can only be set through an invocation. As part of this invocation, the *global* variable x' is inserted into the *local* variable x'. It is important to recognize that x' is an overloaded variable; meaning it is actually two different and distinct variables and its insertion cannot be ignored.

Note that ξ in Equation 6.12 can easily be found using the addition mean equation, which sums the forward and reflected intercept lengths of the ray traveling between the origin and point Q and then divides the sum by two.

We have shown that two of Einstein's Tau invocations return the average intercept time for a ray of light oscillating between the origin (point P) and a point Q on the moving system. Figure 6–8 illustrates the relationship between the ray of light and the position of the moving system when the ray has reached the x axis average intercept length. It is important to notice that when the ray of light has traveled the x axis average intercept length, it has not yet reached the point Q on the moving system. This is an important characteristic that we will revisit when we consider Einstein's second postulate.

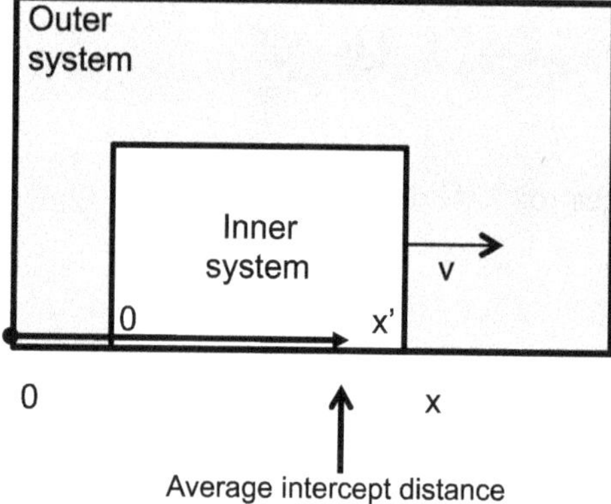

Figure 6–8 Illustration of the ray of light with respect to the inner and outer systems after it has traveled the average intercept length along the x axis. Notice that the ray of light has not yet reached x' of the inner system. The ray of light is at position $(\xi,0,0)$ and has not yet reached point Q of the inner system, which is at $(x' + v\tau_{x'},0,0)$ with respect to the outer system.

The third Tau invocation we will explore is the implicit invocation that occurs when Einstein is developing his y axis transformation. Einstein takes point R on the moving system's y axis, which is at $(0, y, 0)$. This ray of light travels from the origin, P, of the inner system to R. Upon reaching R, it is reflected back to P on the inner system. As illustrated in Figure 6–9 and consistent with the behaviors of a non–nested relationship, the ray of light and the moving system both move with respect to the stationary system.

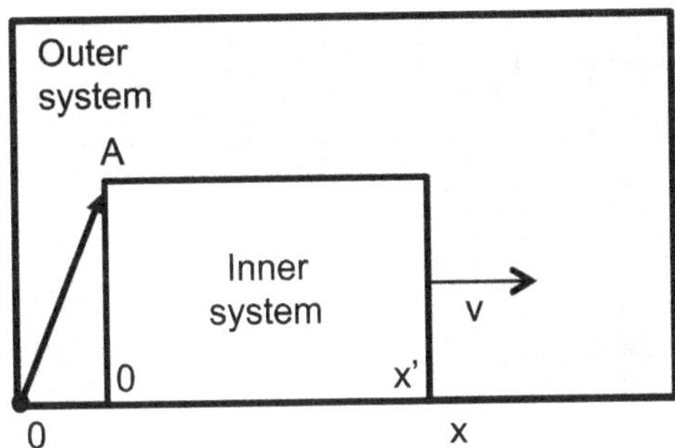

Figure 6–9 The inner system moves forward at velocity v and moves with respect to the outer system using the translation transformation. The ray moves with respect to the outer system along the path from the origin to a point at position $(0, y, 0)$ of the inner system.

We have already shown in Chapter 4 that the forward and reflected intercept time for the oscillating system traveling between the origin and point R on the y axis of the moving system, is:

$$\frac{y}{\sqrt{c^2 - v^2}}$$

To illustrate what is occurring in this invocation, we show it explicitly as:

$$\tau(0,y,0,\frac{y}{\sqrt{c^2-v^2}})$$

where the forward intercept time to point R is used as the fourth argument. The first three arguments represent a point that is at position y on the moving system's y axis. Einstein does not invoke Tau formally. Instead, he informally invokes Tau textually by "inserting" $\frac{y}{\sqrt{c^2-v^2}}$ into t and inserting **0** into x', producing the instantiated expression:

$$\equiv \frac{y}{\sqrt{c^2-v^2}}$$

This is the average intercept time for a ray traveling between the origin and point R on the y axis of the moving system. The average intercept length, η, for this ray is found by multiplying the average intercept time for point R by the velocity of the oscillating system, c, resulting in:

$$\eta = \frac{y}{\sqrt{1-\frac{v^2}{c^2}}} \qquad \text{Eq. 6.13}$$

The fourth Tau invocation for a point S on $(0,0,z)$ of the inner system follows a similar analysis. From our discussion in Chapter 4, we know that the forward intercept time between the origin and S is:

$$\frac{z}{\sqrt{c^2-v^2}}$$

If Tau were explicitly invoked as:

$$\tau(0,0,z,\frac{z}{\sqrt{c^2-v^2}})$$

the first three arguments would represent a point on the z axis of the moving system and the fourth argument would be the forward intercept time to this point. Einstein performs this invocation implicitly by "inserting" $\frac{z}{\sqrt{c^2-v^2}}$ into t and inserting 0 into x', producing the instantiated expression:

$$\equiv \frac{z}{\sqrt{c^2-v^2}}$$

This is the average intercept time for a ray traveling between the origin and point S on the z axis of the moving system. The average intercept length, ς, for this ray is found by multiplying the average intercept time for point S by the velocity of the oscillating system, c, resulting in:

$$\varsigma = \frac{z}{\sqrt{1-\frac{v^2}{c^2}}} \qquad \text{Eq. 6.14}$$

In four of Einstein's six function invocations, we have shown that Tau returns the average intercept time for a ray of light oscillating between points on the moving system. We have also shown that Einstein's transformation equations, prior to his final substitution, are the average intercept lengths to three different points on the moving system.

Einstein's two remaining Tau invocations are interesting because they both use the origin as the point represented by the first three arguments, but both use different values in the fourth argument. This inconsistent usage is a problem because it associates multiple forward intercept times with the same point. Each invocation must be explored to determine whether it is correct; and if incorrect, to determine whether a correction can be offered.

The fifth invocation we will explore is Einstein's explicit invocation:

$$\tau(0,0,0,0)$$

In this case, the ray would oscillate between the origin and the origin. Since no motion occurs in this case, the forward intercept time is zero, which is properly reflected in the forth argument. This invocation's use and result is consistent with the other four we have examined and it correctly returns zero as the average intercept time.

We have shown that in five of Einstein's six invocations that Tau returns the average intercept times to specific points on the moving system. Einstein then uses the returned values to produce the average intercept lengths to these specific points. Notice that each of the average intercept lengths can easily be found using the addition mean equation to divide by two the sum of the forward and reflected intercept lengths of the ray oscillating between the origin and one of the specified points. Each of these invocations is consistent with Tau's use of the subtraction mean equation to return the average intercept time between the origin of the moving system and a point on the moving system.

We have one invocation left to examine, which Einstein explicitly states as:

$$\tau(0,0,0,\frac{x'}{c-v}+\frac{x'}{c+v})$$

We immediately notice that the fourth argument is not the forward intercept time to point P (the origin). Instead, the fourth argument is the total oscillation time from P to Q and back to P because it is the sum of the forward and reflected intercept times. This use of the total oscillating time is inconsistent with Einstein's other five invocations. The fact that Tau produces an answer reveals something important about its behavior: When the first argument is zero, the Tau function simply returns the value of the fourth argument. This inconsistency in the meaning of the fourth argument will produce an error, because the Tau function has no way of knowing whether the fourth argument is the forward intercept time or the total oscillation time.

To correct this problem, Einstein's original statement should be written as:

$$\frac{1}{2}\left[\tau\left(x',0,0,t+\frac{x'}{c-v}\right)+\tau\left(-x',0,0,t+\frac{x'}{c+v}\right)\right]=\tau\left(x',0,0,t+\frac{x'}{c-v}\right)$$

This statement would match Einstein's intent, which was to show that Tau would return the same value for a ray traveling to the left on the x axis as for a ray traveling to the right on the x axis. Because we have shown that Tau returns the average intercept length, this equation would have satisfied Einstein's goal. This equation is consistent because in each of these invocations the fourth argument consistently represents the forward intercept time between the origin and the point represented by the first three arguments. This set of invocations also overcomes the inconsistency in Einstein's work, since a point can only be invoked with one value (eg, the forward intercept time) as the fourth argument. Said another way, using notation consistent with that presented in Chapter 4, the Tau function essentially solves for:

$$\tau_A = \tau_F - \tau_H$$

which says that the average intercept time τ_A is found by subtracting the intercept time half–difference τ_H from the forward intercept time τ_F. We have shown that the intercept time half–different is part of the Tau function and that the forward intercept time is passed as the fourth argument. We have now shown that Tau uses the subtraction mean equation to compute the average intercept time.

We now revisit the simplifying assumption where we set $t = 0$ during each of the explicit invocations. Recall that the t variable used during the function invocation is the global t variable, which is used to represent the time that the moving system has been in motion. When t is set to zero in the function's invocation, we are able to perform the analysis we just discussed and Tau returns the average intercept time. When t is not zero, it represents the time that the moving system has been in motion. Tau then uses that values as the starting time and adds it to the average intercept time. For example, if the moving system has been in motion for 100 seconds and the average intercept time is one second, then Tau will return 101 seconds as the result. For simplicity, Tau should be invoked without adding t to the forward intercept time. If required, it could easily be added to Tau's result, in which case the equation would be written as:

$$t+\frac{1}{2}\left[\tau\left(x',0,0,\frac{x'}{c-v}\right)+\tau\left(-x',0,0,\frac{x'}{c+v}\right)\right]=t+\tau\left(x',0,0,\frac{x'}{c-v}\right)$$

Said another way: *Tau uses the subtraction mean equation to return the average intercept time relative to a starting time of t instead of a starting time of zero.* This is represented as:

$$\tau_{A+t} = t + \tau_F - \tau_H$$
$$= t + \tau_A$$

where t is the amount of time that the moving system has been in motion and τ_A is the average intercept time.

With each of Einstein's function invocations examined, we make one final observation: Einstein's Tau function ignores the second and third parameters entirely. This occurs because he did not develop his PDE with enough unique function invocations to discern their behaviors. It appears that Einstein's Tau function will not work for points that do not lie on one of the moving system's axes. While this may appear to be a problem, Tau will return the proper result as long as the forward intercept time to the desired point is used as the fourth argument. Interestingly, Einstein does not define the average intercept lengths or times associated with the point he explicitly defined; point T, which is at (x', y, z) of the moving system. Said differently, Einstein never addresses the point T that he introduced in his text.

We have now examined each of Einstein's six Tau invocations to reverse engineer the meaning of Tau, along with the meaning of each of its local variables, thus taking the mystery out of its behavior. In a general sense, *Tau uses the subtraction mean equation to return the average intercept time when provided the coordinates of a point on the inner system as the first three arguments and the forward intercept time as the fourth argument.* Tau returns the average intercept time to specific points on the inner system, which Einstein multiplies by c to find three different average intercept lengths.

Using Einstein's variables: ξ, η, and ς, are average intercept lengths for rays of light traveling between the origin and three different points on the inner system. The variable ξ describes the average intercept length of the ray traveling between the origin and point Q at $(x',0,0)$ along the x axis of the inner system. The variable η describes the average intercept length of the ray traveling between the origin and point R at $(0,y,0)$ along the y axis of the inner system. The variable ς describes the average intercept length of the ray traveling between the origin and point S at $(0,0,z)$ along the z axis of the inner system.

Einstein's transformation equations, prior to his final substitution, are the average intercept lengths for a ray of light oscillating between the origin and specific points on the moving system. Contrary to Einstein's statements, they do not define the position of the moving system, which is moving according to the translation transformation.

Because Tau returns an average intercept time and the moving system always moves according to the Newtonian equation, we can easily determine the position of the moving system using the equation:

$$\begin{aligned} x &= x' + v(t + \tau_A) \\ &= x' + v\tau_{A+t} \end{aligned} \qquad \text{Eq. 6.15}$$

where τ_A is the average intercept time between the origin and a given point on the moving system and t is the time that the moving system had already been in motion. Consistent with Einstein's choice of variables, x' represents the beginning position and x represents the position of the moving system after time $t + \tau_A$.

While the equations for ξ, η, and ς discussed in this section appear in Einstein's derivation, these are not Einstein's final transformation equations. To arrive at Einstein's equations, we must now examine the remaining two mathematical operations Einstein performed in his work.

Normalization

Although Einstein does not have their proper meaning, he has arrived at three correct equations for the average intercept lengths from the origin to points Q, R, and S on the inner system:

$$\xi = \frac{x'}{1 - \frac{v^2}{c^2}}$$

$$\eta = \frac{y}{\sqrt{1 - \frac{v^2}{c^2}}} \qquad \text{Eqs. 6.16}$$

$$\varsigma = \frac{z}{\sqrt{1 - \frac{v^2}{c^2}}}$$

To understand what normalization is and why Einstein performed it as part of his derivation, one must remember that he did not recognize that he had already mathematically stated, in multiple places, that the moving system's motion is always found using the Newtonian equation. Additionally, he did not recognize his "transformation equations" as average intercept lengths for three different rays of light oscillating between the origin and three different points on the moving system.

When a moving system is in constant translatory motion along the x axis, its position will change along the x axis. In other words, the x coordinate will change, but the y and z coordinates will remain unchanged. Einstein recognized this as a requirement for his final equations. When examining the average intercept length equations, Equation 6.16, the η and ς values are too large by a factor of $\frac{1}{\sqrt{1 - \frac{v^2}{c^2}}}$.

Einstein's correction was to divide each of the equations by this factor, a step called **normalization**. Normalization will adjust the value of each equation. Specifically, it will adjust η and the value ς so that the transformed y and z values appear unchanged. Einstein takes these **normalized average intercept lengths** and mistreats them as the position of the moving system. This undeclared normalization step occurs when Einstein says: "*Substituting x' with its value.*" Since Einstein has already stated that $x' = x - vt$, this substitution should produce:

$$\xi = \frac{x - vt}{1 - \frac{v^2}{c^2}}$$

$$\eta = \frac{y}{\sqrt{1 - \frac{v^2}{c^2}}} \qquad \text{Eqs. 6.17}$$

$$\varsigma = \frac{z}{\sqrt{1 - \frac{v^2}{c^2}}}$$

where x' was replaced by $x - vt$ alone. Instead, Einstein produces the normalized equations:

$$\xi = \frac{x - vt}{\sqrt{1 - \frac{v^2}{c^2}}}$$

$$\eta = y \qquad \text{Eqs. 6.18}$$

$$\varsigma = z$$

We will use an example to explain how Einstein's normalization step introduces error into his equations. Assume that your best friend borrowed \$25 from you. To represent this loan, your friend writes it in his notebook so that he will remember to pay you back. He writes:

$$b^2 = 25$$

where b^2 is the amount that he borrowed. Later that week, your sister, who is financially astute and helps you manage your money, demands that you charge your friend 10% interest. She sends a note to your friend demanding that he pay you back, with 10% interest. Unaware of the equation in your friend's notebook, your sister sends your friend a note with the equation:

$$p = b + b * 0.1$$

Your friend receives your sister's note and sends you a check for $5.50, as payment in full. You immediately call your friend and ask what happened and want to know when you'll receive the rest of your money. You loaned him $25 and you are only receiving $5.50 in return! Your friend insists that there is no problem because both of the equations he is using are syntactically correct: He wrote a valid math statement in his notebook and your sister sent him a message with a valid math statement. So, what could possibly be wrong?

The problem is that when performing all of the steps together, your friend dropped part of a squared value. It is not a syntax mistake because, when taken separately, each equation is mathematically correct. However, the problem occurs when both equations are associated with one another as part of the same story. One equation uses b^2 to represent the amount borrowed while the other uses b to represent the amount borrowed. This is a semantic mistake, one that mirrors the mistake Einstein made while performing his undeclared normalization step. Technically, since normalization produces new equations, you cannot perform this step and retain their assignments into the variables ξ, η, and ς.

Interestingly, Einstein arrives at a very similar equation as was found by Lorentz. In Einstein's 1905 work, he says that in order for velocity to be viewed as c from the perspective of the moving system, it would need to be viewed from the perspective of the stationary system as $\sqrt{c^2 - v^2}$. Einstein's statement mirrors that made by Lorentz's earlier works regarding the observation that the rays traveling along the y and z axes change on average by a factor of $\frac{1}{\sqrt{c^2 - v^2}}$.

While true, we are also reminded that the ray traveling along the x axis changes on average by a factor of $\frac{1}{1 - \frac{v^2}{c^2}}$.

There is no statement in Einstein's or Lorentz's works that proves that this velocity adjustment for a ray traveling along the y or z axes would take precedence as a "time adjustment" over a ray traveling along the x axis.

In order for Einstein to perform this adjustment, he simply drops a β term when he says: "*Substituting x′ with its value*" to produce his final equations. Interestingly, we will later show that Einstein's spherical wave proof was intended to show that light traveled at the same velocity, c, from the perspective of both the moving and stationary systems. His proof, therefore, contradicts the statement used in his derivation that a ray traveling at velocity c from the perspective of the moving system is traveling at velocity $\sqrt{c^2 - v^2}$ from the perspective of the stationary system.

The non–normalized equations are the same as those used by Modern Mechanics. We mentioned in Chapter 1 that the Modern Mechanics equations perform better than Einstein's normalized versions in the Michelson–Morley experiment. Normalization introduced measurable error in Einstein's equations; error that would not be present had he simply used the average intercept lengths. The amount of error introduced by Einstein's normalization step is dependent on how the equations are used. However, we can imagine a case where we define the result as the difference between the transformed value and the original value. In Modern Mechanics, the change in length, ΔL_M, is the average intercept length minus the segment length, $\Delta L_M = \xi_{xA} - x'$, or:

$$\Delta L_M = \frac{x'}{1 - \frac{v^2}{c^2}} - x' \qquad \text{Eq. 6.19}$$

In relativity theory, this change in length, ΔL_R, is the transformed value minus the original value, $\Delta L_R = \xi - x'$, or:

$$\Delta L_R = \frac{x'}{\sqrt{1 - \frac{v^2}{c^2}}} - x' \qquad \text{Eq. 6.20}$$

Modern Mechanics and relativity are different models and, in practice, they treat x' differently. In analyzing experiments using relativity theory, x' represents the total oscillation length. This is most apparent when using the equations to evaluate experiments where wavelength, λ, is used to represent the length of a complete oscillation. In Modern Mechanics, x' represents the segment length, or half the total oscillation. In other words, when analyzing experiments using Modern Mechanics, x' is $\frac{\lambda}{2}$.

We can now find the amount of error by subtracting the change in length in Modern Mechanics from the change in length in relativity, or $err = \Delta L_{MM} - \Delta L_{RT}$, which is:

$$err = \left(\frac{x'}{1 - \frac{v^2}{c^2}} - x' \right) - \left(\frac{2x'}{\sqrt{1 - \frac{v^2}{c^2}}} - 2x' \right) \qquad \text{Eq. 6.21}$$

The amount of error between these equations is surprisingly small, but measurable. At low velocities, the amount of error is extremely small and it would be hard to differentiate between the two models. In other words, Einstein's normalized equations will provide entirely acceptable results. Depending on the accuracy of the experiment, the error in Einstein's equations can easily be mistaken for experimental error. Note that the amount of error must be individually computed for each experiment and are dependent on how the equations are used and combined.

Relativity uses normalized average intercept times and lengths. These normalized equations introduce error, which is why they do not perform as well as the non–normalized equations used in Modern Mechanics. We will compare the accuracy of Modern Mechanics with relativity in greater detail in Chapter 7.

Incorrectly Simplifying the Tau Equation

As discussed throughout this chapter, functions must be invoked before they can be used. A key implication is that extreme care must be taken when simplifying a function that is comprised of overloaded variables. We now examine Einstein's *simplification* of Tau, where he mistreats the Tau function as if it were an equation.

Einstein writes his Tau function in a manner that allows him and others to easily mistake it for an equation. This leads to two syntax mistakes, both of which occur when he says: "*Substituting x' with its value.*" Notice that:

$$x' = x - vt$$

is the translation equation, where each of the variables exists in the global namespace. However, in the body of Einstein's Tau function, the x' variable is a different variable than appears in this translation equation. To improve clarity, we will refer to the different instances of x' as $global\_x'$ and $local\_x'$. The $global\_x'$ variable is used in the translation equation and the $local\_x'$ variable is used in the Tau function. The two are different variables and cannot be used interchangeably. However, because they are overloaded variables and because Einstein mistreats Tau as an equation, he incorrectly performs a substitution involving the $local\_x'$ variable.

Einstein does not recognize the distinction between functions and equations and how they interact with the various namespaces. Upon *"substituting x' with is value,"* Einstein replaced $local\_x'$ with $global\_x'$. This is a syntax mistake for two reasons. First, the two variables are not the same. Substitutions occur between variables in the same namespace. Second, functions introduce new namespaces and must be invoked to place values into their local variables. This creates an extremely subtle and difficult–to–detect problem in Einstein's work that occurs when he *substitutes* x' in the function body.

Beginning with:

$$\tau = t - \frac{vx'}{c^2 - v^2} \qquad \text{Eq. 6.22}$$

Einstein *"[substitutes] x' with its value"* instead of invoking Tau as a function to incorrectly arrive at:

$$\tau = t - \frac{v(x - vt)}{c^2 - v^2} \qquad \text{Eq. 6.23}$$

Einstein's substitution of the local x' with its value is an incorrect invocation and a syntax mistake. The only way to place values from the global namespace into variables in the function namespace is through a function invocation.

Tau now consists of both overloaded t variables, a fact made more apparent by rewriting the function to make its local and global t variables clear,

$$\tau = \alpha(local\_t - \frac{v(x - v * global\_t}{c^2 - v^2})$$

Because of the overloaded t variable, Einstein cannot simplify the equation as:

$$\tau = \frac{t - \frac{vx}{c^2}}{\sqrt{1 - \frac{v^2}{c^2}}} \qquad \text{Eq. 6.24}$$

following his normalization step. Einstein's mistake is made apparent if we use the equivalent function definition:

$$\tau(k,l,m,n) = \left\{ \alpha(n - \frac{vk}{c^2 - v^2}) \right\}$$

where k, l, and m are the coordinates of the point and n is the forward intercept time. In this case, it is apparent that neither t nor x' can be "substituted."

Mistakes in Einstein's other works

There is a common belief that if a math statement looks right, the statement is correct. This belief does not take into account constraints introduced by earlier statements. This can make it appear that math allows things that are not allowed. For example, consider the apparent proof of 1=0 that was presented earlier in this chapter. While each step is syntactically correct, the proof fails when you examine its semantics. Semantic mistakes are often difficult to identify because every step looks syntactically correct.

Up to this point, we have only examined Einstein's 1905 paper. However, he makes mistakes in his later derivations as well. One key mistake that he repeats in several books and papers is the incorrect use of the transitive relation. The transitive relation is a well–known mathematical relation that says:

$$\text{If } j = k \text{ and } k = l \text{ then } j = l$$

which is mathematically written as:

$$if\ (e_1 = e_2\ and\ e_2 = e_3)\ then\ e_1 = e_3$$

However, there is an important constraint that is overlooked by practitioners unfamiliar with the relation's formal definition:

> *The expressions e_1, e_2, and e_3 must all be members of the same set.*

In other words, they all must be able to take on every value that each of the other expressions can take. For example, if <u>each</u> expression resolves over the **real** set (eg, 1.0, 1.1, 1.2, 1.2 … 2.0, 2.1, 2.2 … 3.0, 3.1, 3.2 …), then this rule can be safely applied. Similarly, if each of the expressions all resolve over the **integer** set (eg, 1, 2, 3 …), then the rule can be safely applied. However, if two of the expressions resolve over the **real** set and the third expression resolves over the **integer** set, then this relation cannot be used. This constraint is often overlooked and can only be found by reviewing the relation's definition using formal math notation.

To understand the transitive relation in formal notation, we first expand the original statement to apply to three common operators and write it as

$$if\ (e_1 Re_2\ and\ e_2 Re_3)\ then\ e_1 Re_3$$

where R is a relational operator ">", "<", or "=". While the transitive relation is informally written in this manner, this statement is incomplete. In formal mathematical notation, the transitive relation is defined as

$$\forall e_1, e_2, e_3 \in S : (e_1 Re_2 \wedge e_2 Re_3) \Rightarrow e_1 Re_3$$

While this statement may look funny to some readers, the expression "$(e_1 Re_2 \wedge e_2 Re_3) \Rightarrow e_1 Re_3$" is formal mathematical notation for "if $j = k$ and $k = l$ then $j = l$." This is where most people focus their attention. However, it is the expression "$\forall e_1, e_2, e_3 \in S :$" that is of more importance, because it defines when the relation can be used. Translated from formal math notation, this statement says each expression e_1, e_2, and e_3 has to be able to express every value that can be expressed by each of the other two expressions. In other words, you cannot use the transitive relation with a constant as one of the expressions unless all of the expressions are only capable of producing that constant alone.

Einstein failed to recognize this constraint, resulting in the misuse of the transitive relation as part of his derivations. Using Einstein's choice of variables from his later works, he begins with the equation for spherical waves in both systems, which he writes as:

$$x^2 + y^2 + z^2 - c^2t^2 = 0$$

and

$$x'^2 + y'^2 + z'^2 - c^2t'^2 = 0$$

To use the transitive relations, Einstein needs three expressions:

1. $x^2 + y^2 + z^2 - c^2t^2$ is expression e_1

2. 0 is expression e_2 and

3. $x'^2 + y'^2 + z'^2 - c^2t'^2$ is expression e_3

Notice in the first expression that there are various combinations of x, y, z, and t that allow expression e_1 to have non–zero values. The same occurs with the variables used in expression e_3. However, expression e_2 is a constant that can only represent one value: 0. The requirement for using the translation relation has not been satisfied and it cannot be used to associate expression e_1 with expression e_3. In other words, Einstein makes a mistake when he joins these statements together using the transitive relation to produce:

$$x^2 + y^2 + z^2 - c^2t^2 = x'^2 + y'^2 + z'^2 - c^2t'^2 \qquad \text{Eq. 6.25}$$

For Einstein to correctly arrive at Equation 6.25 by applying the transitive relation, he would need to replace the constant, zero, with an expression that can hold the same values as the other expressions. This would be possible if the constant zero were replaced with a variable, as in:

$$x^2 + y^2 + z^2 - c^2t^2 = g$$

and

$$x'^2 + y'^2 + z'^2 - c^2t'^2 = g$$

While this use of a variable as the second expression would allow Einstein to use the transitive relation, its use is problematic for him because the two equations would no longer represent equations for spherical waves.

Understanding Relativity's Key Concepts

While we have explained Einstein's equations and derivation, we must show how he combines these equations with his remaining postulate to produce his theory. It is essential that you remember these two facts in order to understand how Einstein developed the conceptual framework for relativity theory.

1. *Einstein does not recognize that the motion of the moving system is always governed by the translation equation.*

2. *Einstein does not recognize the equations as average intercept lengths to three different points and mistreats them as representing the position of the moving system.*

Conceptually, Einstein's theory is based on two principles: the principle of relativity and the principle of the constancy of the velocity of light. We have explored the role of the principle of the constancy of the velocity of light in Einstein's theory. Essentially, this postulate defines a non–nested relationship involving a ray of light, a moving system, and a stationary system. It is also the foundational assumption that allows Einstein to use the subtraction mean equation in his Tau function. Ultimately, it is this assumption that allows him to arrive at the three average intercept lengths he then normalizes to produces his final equations.

Einstein's second principle, the principle of relativity, says that if things behave a certain way when a system is stationary, those things will behave the same way when the system is in motion.

This also means that if something behaves a certain way in the stationary system, then it will behave the same way in the moving system. Intuitively this definition makes sense, as does the behavior. Imagine being seated in the first row of an airplane parked at the gate getting ready to fly across the country. You can sit and talk to the passenger seated next to you, walk up and down the aisle, and even play catch with the person sitting in the last row of the plane. After the plane takes off, you realize that everything you could do while the plane was stationary, you can also do while the plane is cruising at 500*mph* at 35,000 feet. While inaccurate, it is reasonable to conclude that the behavior of electromagnetic forces and optics must be in accordance with this principle. Einstein's principle of relativity essentially defines a nested relationship. Using Einstein's words, "*the same laws of electrodynamics and optics will be valid for all frames of reference,*" which is formalized in the principle of relativity as:

> *If two coordinate systems are in uniform translatory motion relative to each other, the laws according to which the states of a physical system change do not depend on which of the two systems these changes are related to.*

Einstein's principle of relativity says that the same behaviors that occurred when the system was stationary will occur when the system is in motion. There are three important implications of this principle:

1. *The velocity of a ray of light traveling in any direction in each system must be* c.

2. *If a ray of light travels a length* x' *in time* t *in one direction, then it must behave the same way in all directions, regardless of whether the system is stationary or in motion.*

3. *Following from the above implications, the time, t, required for the ray of light to travel the length of each segment must be* $\frac{x'}{c}$.

While Einstein's principle of relativity is compatible with the Modern Mechanics behavior of a nested relationship, combining it with the principle of the constancy of the velocity of light is problematic. Einstein's principle of the constancy of the velocity of light is associated with a non–nested relationship, while his principle of relativity is associated with a nested relationship. Conceptually, Einstein has only one type of system relationship, which requires that he associate his two concepts using the spherical wave proof.

Modern Mechanics defined two phrases: "with respect to" and "from the perspective of." When we describe something *with respect to*, we generally use absolute measurements. When we use *from the perspective of*, we will use relative measurements. For example, if a ray of light is moving at c with respect to the stationary system, then that is its absolute velocity. If the moving system is moving at velocity v, then from the perspective of the moving system, the velocity of the ray as it travels the length of the forward segment is $c - v$. Similarly, from the perspective of the moving system, the velocity of the ray of light as it travels the length of the reflected segment is $c + v$.

According to Einstein's principle of relativity, the velocity of the ray of light as measured from the perspective of the moving system is not $c - v$ and $c + v$, but must always be c. This is an apparent contradiction, but can be explained using a three–system model as an example. Assume a bus (moving system) is on a straight road (stationary system). Painted on the road are ruler marks at $1m$ intervals spanning the length of the road. Painted on the floor of the bus are ruler marks at $1m$ intervals spanning the length of the bus. The length of the bus is x'. Now imagine a man (oscillating system) who will walk at a constant velocity from the rear of the bus to the front of the bus, where upon reaching the front of the bus he will reverse direction and walk at the same constant velocity until he reaches the rear of the bus. We must now explain the man's behavior using the ruler on the bus. As per convention throughout this book, the rear of the bus is labeled P and the front of the bus is labeled Q. We will not use the normalized values in order to better illustrate what is occurring.

Imagine you have a watch, which you will use to determine how long it takes the man to walk from P to Q and return to P. Assume that when the bus is stationary it takes the man 15 seconds to complete this oscillation: 7.5 seconds in each direction. If the man were tireless, he could repeat these oscillations indefinitely, allowing you to leave your watch at home because you know that when the man has completed four oscillations, exactly one minute has passed.

Because we were not specific about the man's location, let's first place him on the bus, forming a nested relationship. In this case, when the man has completed four oscillations, exactly one minute has passed. This is true regardless of how fast the bus might be moving. It is true for anyone who is able to measure time using their wristwatch, regardless of whether they are on the bus or on the road. This is what is expected and is consistent with concepts associated with the principle of relativity. However, it is not consistent with Einstein's other postulate, because it is not consistent with a non–nested relationship that says that the oscillating system moves with respect to the stationary system.

We now place the man on the road and again ask him to perform his oscillations. With the man on the road and when the bus is stationary, the man completes each oscillation in 15 seconds. Once again, four oscillations represent one minute. However, when the bus is moving forward at velocity v, we must describe the man's behavior using only the ruler on the bus because we no longer have the wristwatch. As previously discussed, when the man has completed four oscillations we presume that one minute has elapsed. This is our means of measuring time, because we no longer have our watch. Using the bus as the ruler makes reasonable sense, because the man is moving from P to Q covering a distance of x'. So, it makes conceptual sense, when using relative measurements, that if he is moving x' in one direction and x' in the other direction, his total time would be 15 seconds. Unfortunately, while this is what is occurring when computing time using measurements from the perspective of the bus, it is not what is actually occurring because the man is moving with respect to the road. As illustrated in Figure 6–10, he has walked the forward intercept length in one direction and the reflected intercept length in the other direction.

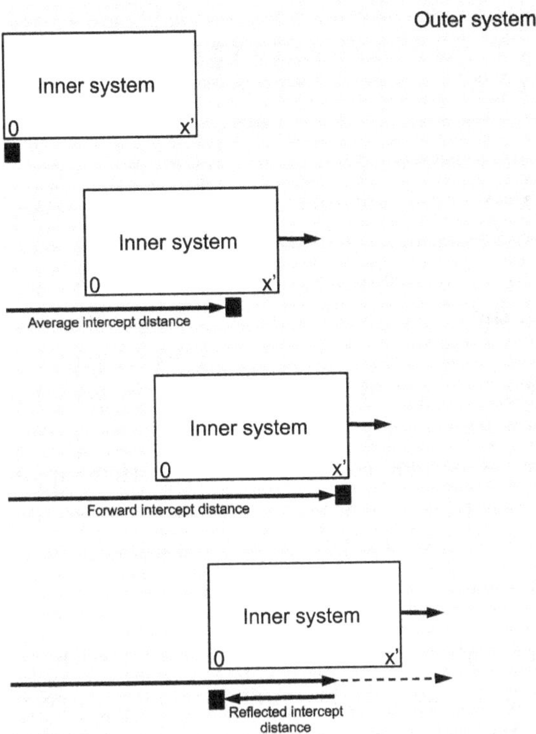

Figure 6–10 The positions of the inner system (eg, bus) and oscillating system (eg, man).

At this point, Einstein has a lot of equations and concepts that appear to conflict with one another. His goal, however, is to show that his principles and math are consistent. According to the principle of relativity, since the man took 15 seconds to travel the length x' of the bus when it was stationary, **this *must* also be the case when the bus is in motion**, and is the case if we measure a unit of time as being the oscillations of the man. In fact, because we no longer have our watch on our wrist and the man's oscillations are the only way that we are keeping time, then in relative terms, each oscillation is 15 seconds. We know, however, that the man is actually walking on the road and is traveling the length of the forward and reflected intercept lengths to complete his oscillations; requiring a total amount of time that is the sum of the forward and reflected intercept times. Because we are using the man as our timekeeping device, time appears to run slower because one oscillation now takes longer than the 15 seconds it took when the bus was stationary. Because time is measured by the man's oscillations (not his wristwatch), and we have defined each oscillation to mean 15 seconds, time runs slower when the bus is in motion. Because it now takes longer for the man to complete four oscillations, anyone using the number of completed oscillations to keep time will observe time running slower. The faster the bus moves, the longer it takes to complete each oscillation; which means the slower time appears to run. This slowing of time is what Einstein defines as "time dilation." Of course, no problem would exist with timekeeping if we had our watch on our wrist.

The man has walked the forward intercept length, but from the perspective of the bus, he has only walked x'. According to the principle of relativity, the man has walked x' regardless of the velocity of the moving system. But we know that the man has walked the length of forward intercept length, because he moves with respect to the road. Importantly, Einstein does not use the forward intercept length to mean that the man has reached the front of the bus, but instead uses the average intercept length. He then attempts to reconcile the relationship between x' and the average intercept length by concluding that when the moving system (eg, the man) has traveled a distance of x' from its perspective, it has actually traveled the average intercept length from the perspective of the stationary system. Einstein names this relationship "length contraction." In this case, Einstein presumes that the average intercept length is measured with respect to the stationary system. While conceptually interesting, it is incorrect because the size of the bus never changes and the position of the bus is not dependent on how far the man walks. It is also important to recognize Einstein's use of the average intercept length means that when the man has traveled the average intercept length, he has not yet reached the front of the bus. As illustrated in Figure 6–10, when the bus is moving, the average intercept length will be less than the forward intercept length, which is the distance required to arrive at x'.

Einstein's interpretation suffers from two problems. First, it fails to recognize the actual distances and positions associated with the moving and oscillating systems, as shown in Figure 6–10. The moving system's position is determined by the translation equations and is not dependent on a length contraction relationship. Second, it mistakes measurements associated with the oscillating system (eg, the man) with measurements associated with the inner system (eg, the bus).

Einstein uses the average intercept length as an element of his work. If, instead of the average intercept length, he used the forward and reflected intercept lengths, we would immediately identify their lengths as being different, which would violate the principle of relativity. While his use of the average intercept length remains consistent with this principle, he must provide an explanation for the forward and reflected intercept lengths. Specifically, what Einstein must now do is show that from the perspective of the moving system, when the man has traveled the average intercept length, he will arrive at Q.

Einstein correctly says that the forward and reflected intercept lengths are measured with respect to the stationary system, but in this case incorrectly concludes that the average intercept length is measured with respect to the moving system. In actuality, the forward, reflected, and average intercepts are each measured with respect to the stationary system in a non–nested relationship. This creates an interesting situation where a man walks the forward intercept length from the perspective of the stationary system, but is presumed to have walked the average intercept length from the perspective of the moving system.

Similarly, he walks in the opposite direction the reflected intercept length from the perspective of the stationary system, but is presumed to have walked the average intercept length from the perspective of the moving system.

Einstein's incorrect use of *from the perspective of* and *with respect to* enables him to now say that the distance the man walks in each direction is different from the perspective of the stationary system (eg, forward intercept length versus the reflected intercept length), but from the perspective of the moving system they are the same (eg, average intercept length in both directions). This is a critical concept in Einstein's work, one that he names "simultaneity."

Time dilation, length contraction, and simultaneity are all critical terms used to explain the various mathematical relationships in Einstein's work. Some people believe that time dilation and length contraction have been experimentally confirmed. In actuality what has been observed is a change in wavelength and a change in frequency. In fact, the proper treatment of types requires distinguishing between wavelength and length, and between frequency and time. Wavelength is not the same as length and frequency is not the same as inversed time. Unfortunately, this type of mistreatment is common.

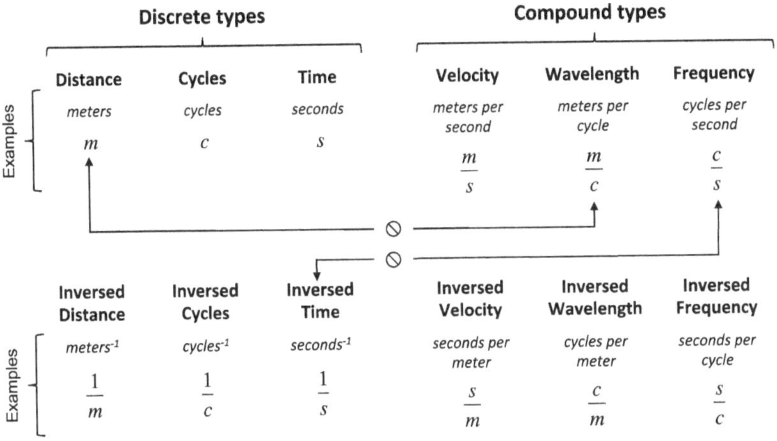

Figure 6–11 Incorrect type relationships. Contrary to how wavelength and frequency are presented in some textbooks, wavelength is a compound type and must not be confused with length, which is a discrete type. Similarly, frequency is a compound type that must not be confused with inverse time, which is a discrete type.

Most mistakes associated with type management do not result in catastrophic failures. As a result, many scientists consistently make this kind of mistake without ever realizing that it is a problem. Additionally, equations involving improper types often produce the correct quantity, masking the underlying type problem.

Einstein requires the spherical wave proof because it is what would have proved that when a ray of light had traveled the average intercept length, it would have arrived at the reflection point. Specifically, Einstein needed to show that the distance and amount of time for the man to travel a specific length in one direction would be the same as when the man travels in any direction, since this would be the case when the bus was stationary. He not only needed to show this for points on the x axis, he needed to show it for any point. Said another way,

Einstein must show that when something has traveled the average intercept length in any direction, it has arrived at the reflection point, regardless of whether the system is stationary or in motion.

Specifically, if a spherical light wave were emitted at time zero, and an infinite amount of sensors where placed n units away from the source, the rays of light would hit each sensor at the same time. The principle of relativity says this behavior must be true regardless of the velocity of the moving system (assuming the velocity of the moving system is less than the velocity of the ray of light). So, Einstein needed to show that when the moving system was in motion, a valid light bubble would be formed from the perspective of the moving system. We have already shown in Chapter 1 that his proof has failed.

We have shown why Einstein's theory required the spherical wave proof. We have also shown how the successful completion of the proof would result in time dilation, length contraction and simultaneity. This discussion also highlights why counterarguments suggesting that Einstein's proof only applies to "a single point" or that his equations are the only things that matter are nonsensical. A proof based solely on a specific point, or adherence only to the equations, would not prove the compatibility of Einstein's two postulates. Einstein's proof failed and we have explained the reasons for the failure: His work only gives normalized average intercept length equations to three points and an x axis normalized average intercept time equation.

Now we examine the equations associated with length contraction and time dilation. Begin with Einstein's normalized average intercept length along the x axis:

$$\xi = \frac{x'}{\sqrt{1 - \frac{v^2}{c^2}}}$$

Since according to relativity, $\frac{x'}{c}=t$ must be true from the perspective of the stationary system, the equation is divided by c to produce the normalized average intercept time as:

$$\tau=\frac{x'}{c\sqrt{1-\frac{v^2}{c^2}}}$$

which is rewritten as:

$$\tau=\frac{t}{\sqrt{1-\frac{v^2}{c^2}}}$$

This is called time dilation because the relative measurement of time will make it appear slower, or dilated, compared with that measured in the stationary system. As previously discussed, the best way to think of time dilation is to realize that time is measured by counting the oscillations of the ray of light. In other words, the ray of light is the clock and when both systems are in motion, it is obvious that the clock runs slower, because it takes longer for the ray to complete each oscillation. This equation presumes that time is measured by counting the oscillations along the x axis. There is no reason that oscillations between the origin and Q should take precedence for measuring time over any other points on the moving system. Time could have easily been measured by counting the oscillations between the origin and R, S, or T. In addition, time could be measured by another means, one that is not associated with the motion of the moving system.

Several alternative derivations of the relativity equations, including Einstein's own work, use this equation as a starting point; failing to recognize that this equation *is the result of* the relativity derivation, not a statement of fact. Any equation that uses this time equation as part of its derivation has automatically assumed normalization and time dilation. Such use will always allow one to arrive incorrectly at Einstein's final normalized equations.

Summary

This chapter evaluated and analyzed Einstein's work, revealing the correct meaning of his equations and explaining the purpose and use of his Tau function. Relativity theory consists of mathematical steps and assumptions that uniquely combine to produce equations that are remarkably good. They perform better than the classical mechanics equations because they provide scaling transformations to explain different things. The classical mechanics equations explain the motion of a moving system. The Modern Mechanics equations, and Einstein's normalized versions of those equations, are scaling transformations that explain oscillation in moving systems. While they can be useful, Einstein's normalization step introduces error, which is not present in the Modern Mechanics equations.

Einstein's equations suffer from three problems: First, relativity theory is not an accurate description of the equations. Einstein found the average intercept times and lengths to three different points on the moving system. Second, he failed to recognize that the translation transformation always applies to the moving system and incorrectly normalized the average intercept lengths to represent the position of the moving system. Third, due to Einstein's undisclosed normalization step, his final equations are less accurate than those associated with Modern Mechanics. None of the problems identified with Einstein's work can be corrected, because relativity theory is fundamentally a two–system theory that attempts to explain three–system behavior.

Missing from Einstein's work is an analysis of the behavior of Tau. While he assumed that Tau represented time in the moving system, Tau actually returns the average intercept time of an oscillating ray of light in a non–nested relationship. Tau takes four arguments, the first three arguments represent the coordinates of the point on the moving system and the fourth argument represents the forward intercept time from the moving system's origin to that point. Tau was one of the least–understood aspects of Einstein's work. However, we now understand the meaning of the function body, the function arguments, and the intercept time half–difference in the function body; Tau returns the average intercept time using the subtraction mean equation.

Relativity theory is unique because it is a combination of ideas and mistakes that produced a set of equations that worked better than its classical mechanics alternative. Relativity, while wrong, worked well for a very long time. However, as we have discussed, not only does Einstein's spherical wave proof fail, we have shown exactly where and why Einstein's theory falls apart. In addition, we have introduced how his theory can be wrong and yet produce accurate results. In a nutshell, Einstein found equations that explained something different than what he thought.

Modern Mechanics, in comparison, does not use Einstein's assumptions or equations. It is derived from the foundations of geometric transformations. It differs from classical mechanics by using a third system to explain oscillation, as well as the explicit use of average intercept lengths and times. Because Modern Mechanics does not normalize its equations, it produces better results than the relativity equations. Unlike relativity, Modern Mechanics is not limited to describing electromagnetic force and optics. It supports particle and wave mediums that can travel or propagate faster than light, and does not require non–intuitive concepts such as time dilation and length contraction.

Chapter 7 Measurement and Experiment Analysis

Modern Mechanics is extremely powerful. It is developed on the foundation of geometric transformations, making it intuitive and easy to understand. While Modern Mechanics is mathematically and conceptually sound, it cannot replace any leading theory on theoretical grounds alone. Theories do not exist in isolation. They are one half of an important relationship. The other half is experimental observation and analysis.

Theories make predictions that experiments either support or disprove. Experiments are extremely powerful and have the power to include or eliminate theories as possible explanations of the world around us. One could argue that experiments are more powerful than theories, because if an experiment produces a result that conflicts with a theoretical expectation, and that experiment is deemed sound, then the theory is called into question and revised, or discarded entirely.

Classical mechanics was the leading theory of motion until the early 20th century, when it failed to provide an explanation for experiments involving the electromagnetic force. Specifically, classical mechanics did not effectively predict or explain the results of the Michelson–Morley interferometer experiment. Introduced in Chapter 1, the Michelson–Morley experiment is one of the most important experiments associated with the electromagnetic force. Its results dismissed many of the leading theories of its day and left a gap that required a new explanation. Any theory intended to serve as a unified model must explain the Michelson–Morley experimental result. Because classical mechanics is unable to explain the Michelson–Morley experiment, it cannot serve as a unified theory.

Einstein's theory of relativity filled the theoretical gap introduced by the Michelson–Morley experimental result. It introduces its own terminology, concepts, and ideas to explain motion. Relativity theory appears to work well and often provides useful mathematical answers. Although we have shown relativity theory to be invalid, we have yet to address the experimental evidence that appears to support Einstein's work. *How can an invalid theory work well and provide useful answers?* As you will read later in this chapter, some of these acceptable results are due to the similarity in predictions between Modern Mechanics and relativity theory. In other cases experiments long thought to support relativity theory, in reality do not. Similar to the Greek model of the solar system, a theory can provide helpful mathematical answers, but still be mathematically and conceptually incorrect. Since we have already explained the mathematical failure of Einstein's theory, it cannot serve as a unified model, regardless of how well is appears to perform.

While we have explained specific mistakes in Einstein's work, many questions remain. First, how can relativity theory be wrong when it appears to be supported by several foundational experiments like the Michelson–Morley and the Ives–Stilwell experiments? Second, does Modern Mechanics perform as well as, if not better than relativity theory? Third, does Modern Mechanics explain experiments and observations that are currently explained by classical mechanics and quantum mechanics? The answers to these questions will show that Modern Mechanics' predictive and explanatory capabilities match or exceed those of the alternatives: classical mechanics, relativity theory, and quantum mechanics.

The objective of this chapter is straightforward: to show that Modern Mechanics makes accurate predictions and provides answers for experiments that are commonly associated with classical mechanics, quantum mechanics, and relativity theory. We will examine and analyze four experiments and observations: the Doppler effect, which is most commonly associated with classical mechanics; the Michelson–Morley experiment, which is associated with relativity theory; the Ives–Stilwell experiment, which tests Einstein's time dilation and energy equations; and the Young double–slit experiment, which is associated with quantum mechanics. Demonstrating that Modern Mechanics explains the experiments associated with each of these theories is an essential step to demonstrate its value, positioning it as a unified model.

The Doppler Effect

At the turn of the 20th century, scientists were investigating a newly discovered force, called the electromagnetic force. Our current understanding of the electromagnetic force exceeds what was known a century ago. Today we commonly use electricity, computers, the Internet, and wireless communications, each of which is made possible because of our understanding of the electromagnetic force.

Many observations associated with the electromagnetic force are described using the Doppler effect. Named after Austrian physicist Christian Doppler who documented the effect in 1842, the classical mechanics–based Doppler effect explains the apparent change in wavelength resulting from the motion of a system. Apparent change refers to the fact that the wavelength observed by the receiver of a signal may be different from the wavelength that was sent. The Doppler effect is extremely useful. It is used extensively in a variety of applications, ranging from sound and acoustics, satellite communications, medical instrumentation, and radar, just to name a few.

The Doppler effect, also called Doppler shift, defines mathematical equations that explain the *apparent change* in wavelength resulting from the motion of the source, the receiver, the transmission medium, or a combination of the three. To simplify this discussion, we will only discuss cases involving moving sources traveling along the same linear path as a stationary receiver. The Doppler effect can be described in terms of wavelength, frequency, or both. Due to the one–to–one relationship between wavelength and frequency, we use both terms interchangeably, as necessary to help with comprehension.

To illustrate how the Doppler effect works, imagine that you are standing next to some train tracks while the engineer of an approaching train is blowing the locomotive's horn. Since the train is approaching your position, you will hear the horn at a higher frequency than if the train were stationary. Once the train passes and begins moving away from your position, you hear the horn at a lower frequency than if the train were stationary. While the sound you hear changes, the sound that the engineer on the train hears remains unchanged. The source frequency of the horn did not change. However, the received frequency – the sound you hear – appears to change, depending on whether the locomotive was moving toward or away from you. As in real life, if the train is always in motion, you will only hear the horn's *apparent frequency*; you may never hear the horn at its transmission frequency (ie, the frequency that the engineer hears).

The Doppler equations describe the apparent wavelength of the sound you hear in mathematical terms. When an object (eg, the train) moves toward a stationary object (eg, you) the apparent wavelength shortens and the apparent frequency increases. When an object moves away from a stationary object, the apparent wavelength lengthens and the apparent frequency decreases.

While the stationary and moving objects can each serve as the transmitter or receiver, the equations we will develop assume that the moving object is the transmitter and that the stationary object is the receiver. The two most prominent Doppler equations are:

$$\lambda_P = \frac{\lambda_Q}{1 - \frac{v}{w}} \qquad \text{Eq. 7.1}$$

and

$$\lambda_P = \frac{\lambda_Q}{1 + \frac{v}{w}} \qquad \text{Eq. 7.2}$$

where λ_P is the observed wavelength at the stationary object, λ_Q is the original wavelength at the moving object, v is the velocity of the moving system, and w is the propagation speed of the wave. In Equation 7.1, the moving object is traveling away from the stationary object, while in Equation 7.2 the moving object is traveling toward the stationary object. When we consider the specific case of electromagnetic force in a vacuum, the propagation speed of the transmission medium w is replaced by the speed of light, c. In astronomy, where the velocity of the wave is c, equations 7.1 and 7.2 are known as the *blue–shift* and *red–shift equations*.

Intuitively, since Modern Mechanics and classical mechanics are both based on geometric transformations, Modern Mechanics can explain the same things as classical mechanics. Because the relationship between Modern Mechanics and classical mechanics was explained in Chapter 2, it is easy to understand why both will produce the same Doppler equations. Notice that equations 7.1 and 7.2 are almost identical to equations 4.28 and 4.29, with the most visible difference being the replacement of the variable x with λ_Q and the replacement of the instance variables with λ_P. On the surface, one might think we can simply perform these replacements and use the revised equations from Chapter 4 as the Doppler shift equations. While they will consistently produce the correct numeric quantity, they are not the same as equations 7.1 and 7.2. The difference between the equations is due to an extremely subtle but important characteristic called *types*.

Introduced in Chapter 2, types describe variables. They can be discrete or compound. In Chapter 4, the variable x represents a length. As discussed in Chapter 2, a length is a discrete type. On the other hand, equations 7.1 and 7.2 use the variable λ_Q to represent a wavelength. Wavelength is a compound type. Although the equations may appear similar, the difference in the type of the variables used in the numerator makes them different. Technically, the equations from Chapter 4 cannot be used as the Doppler equations due to a mismatch in types. The Doppler equations 7.1 and 7.2 describe behaviors associated with wavelength, while the equations from Chapter 4 describe behaviors associated with length. This extremely subtle distinction is all too easy to overlook. Effective type management is crucial and required to properly adapt equations 4.28 and 4.29 to serve as Doppler equations. Specifically, this adaptation is performed using a technique called **type conversion**. Type conversions are common in computer science where variables of one type must be properly converted into variables of another type. In some cases, the type of the variable changes, but the underlying value remains the same. This often happens when working with length and wavelength. In other cases, both the type and the value change, an example of which occurs when we convert 98.6 degrees Fahrenheit into 37 degrees Celsius. Here, the value changes from 98.6 to 37 and the type changes from Fahrenheit to Celsius.

To correctly use the functions and equations developed in Chapter 4, we must first convert the wavelength λ_Q into a length. Then, the answer from those equations must be converted from a length back into a wavelength that can be assigned to λ_P. To perform these conversions, assume the existence of two conversion functions: *Length()* and *Wavelength()*. During invocation, both functions take one argument. The *Length()* function takes a wavelength argument and returns a length. The *Wavelength()* function takes a length argument and returns a wavelength. With the aid of these conversion functions, we define a new function called the **intercept wavelength function**, which is defined as:

$$\lambda_s(\lambda, v, s, w) = \{Wavelength(\ L(\ Length(\lambda), 0, 0, v, s, w)\ )\} \qquad \text{Fn. 7.1}$$

where λ is the source transmission wavelength, v is the velocity of the moving object, s is the selector, w is the propagation speed of the wave, λ_s is the function name, and *L()* is the intercept length function that was developed in Chapter 4. When the selector is 1, the intercept wavelength function returns the forward intercept wavelength; when the selector is –1, the function returns the reflected intercept wavelength; and when the selector is 0, the function returns the average intercept wavelength.

Instantiation of the intercept wavelength function with $s = 1$ and $s = -1$ will produce the forward and reflected wavelength equations described by equations 7.1 and 7.2, which when written in Modern Mechanics notation are:

$$\lambda_F = \frac{\lambda}{1 - \frac{v}{w}} \qquad \text{Eq. 7.3}$$

and

$$\lambda_R = \frac{\lambda}{1 + \frac{v}{w}} \qquad \text{Eq. 7.4}$$

where v is the velocity of the moving system, w is the propagation speed of the wave through the transmission medium, λ is the wavelength emitted by the moving object, λ_F is the apparent wavelength observed by the stationary object when the moving object is moving away, and λ_R is the apparent wavelength observed by the stationary object when the moving object is moving closer.

Because of the one–to–one relationship between frequency and wavelength, the intercept wavelength function can be used to define the **intercept frequency function** as:

$$f_s(\lambda, v, s, w) = \left\{ \frac{w}{\lambda_s(\lambda, v, s, w)} \right\} \qquad \text{Fn. 7.2}$$

which can also be expressed using instance variables. When the intercept wavelength is expressed as the instance variable $\lambda' = \lambda_s(\lambda, v, s, w)$, the intercept frequency equation can be expressed as the equation:

$$f' = \frac{w}{\lambda'} \qquad \text{Eq. 7.5}$$

where f' is the intercept frequency, λ' is the intercept wavelength, and w is the propagation velocity of the wave through the medium.

We have demonstrated that Modern Mechanics will yield exactly the same equations as the classical mechanics–based Doppler equations. We have already shown how Modern Mechanics and classical mechanics are based on geometric transformations and use the same translation equations. As a result, Modern Mechanics can explain the same experiments and observations that are effectively explained with classical mechanics.

When multiple Doppler equations are combined with one another, proper type management is essential. The misuse of equations for one type and variables of another will lead to incorrect answers. This will be discussed further when we examine the Michelson–Morley experiment.

The Luminiferous Aether and the Michelson–Morley Experiment

In the 19th century, our understanding of light, electronics, and the electromagnetic force was just developing. Many concepts that we currently take for granted had yet to be developed. As scientists started to unlock the mysteries of this new force, they developed theories to explain their observations, and experiments to test their ideas and theories. Many 19th century scientists believed that light traveled through an electromagnetic medium, which they called the luminiferous aether, or simply aether or ether. Interestingly, they did not debate whether an aether existed. Instead they debated whether the aether was completely or partially dragged. They developed two leading, but competing ideas: a partially dragged aether and a completely dragged aether.

The partially dragged aether model and the completely dragged aether model both attempt to answer an important question about the nature of the aether: *Does the aether move with respect to the Earth (inner system), the Sun (outer system from the perspective of the Earth), or the Universe (outer system from the perspective of the Earth and the Sun)?* In many ways, these aether hypotheses share similarities with the ideas and concepts developed in Modern Mechanics. We will use these similarities to explain these ideas using Modern Mechanics terminology.

As discussed earlier, the aether moves with respect to the inner system in a nested system relationship. This behavior describes the motion of the aether if it moves with respect to the Earth. In a non–nested system relationship, the aether moves with respect to the outer system. This behavior describes the motion of the aether if it moves with respect to something other than the Earth, like the Sun or the Universe. Conceptually, a nested system relationship is similar to a completely dragged model and a non–nested system relationship is similar to a partially dragged model. Early experiments involving the electromagnetic force did not clearly support a completely dragged model. This left the partially dragged model as the leading explanation of the luminiferous aether.

Several scientists tested various theories based on a partially dragged aether model. Two of these scientists were Michelson and Morley, who in 1887 published a paper in the American Journal of Science entitled: *On the Relative Motion of the Earth and the Luminiferous Ether*. As mentioned in Chapter 1, their paper describes an experiment that measures the Earth's orbital velocity. If successful, it would confirm a partially dragged aether–based theory proposed by Fresnel, support an electromagnetic aether, and align with classical mechanics. On the other hand, if their experiment was unsuccessful, we would be left without a viable aether–based model to explain the electromagnetic force.

The Michelson–Morley experiment is an ingenious test, designed to measure the change in wavelength of light waves as the Earth orbits the Sun. The experiment is based on the idea that if light travels as a wave, it would behave according to the classical mechanics–based equations associated with Doppler shifts. Specially, it would behave like a non–nested system relationship with the Sun serving as the outer system. They believed that the change in wavelength could be observed and measured, and those measurements could be used to calculate the Earth's orbital velocity.

As mentioned in Chapter 1, the Michelson–Morley experiment did not produce the results they expected. Michelson and Morley use their data with their equations to calculate an Earth orbital velocity of $8km/s$. This was not close to the $30km/s$ they expected. Since their expected result presumes that the aether travels with respect to the Sun and they knew that the Earth orbits the Sun at $30km/s$, they expect their measurements to confirm this velocity. Instead, what they conclude is:

> *"It appears, from all that precedes, reasonably certain that if there be any relative motion between the Earth and the Luminiferous aether, it must be quite small; quite small enough to entirely refute Fresnel's [partially dragged aether] explanation of the aberration. Stokes has given a [completely dragged aether] theory of aberration which assumes the aether at the Earth's surface to be at rest with regard to the latter, and only required in addition that the relative velocity have potential; but Lorentz shows that these conditions are incompatible."*

Michelson and Morley concluded that the partially dragged aether model proposed by Fresnel and the completely dragged aether model proposed by Stokes both fail. They continue and conclude that another aether–based model proposed by Lorentz also fails.

The Michelson–Morley experiment is one of the most important experiments associated with modern science, because it fails to provide support for an electromagnetic aether. Said another way, their experiment appears to completely disprove the idea of a luminiferous aether. Because researchers, scientists, and mathematicians could not find anything wrong with their work, scientists had to accept their results and conclusion. This left a theoretical void that would eventually be filled by Einstein's relativity theory.

In 1905, 18 years after the publication of the Michelson–Morley paper, Einstein published his now–famous paper that established relativity theory. Relativity theory was important because it filled a gap by providing a mathematical model and theoretical explanation of the electromagnetic force. Interestingly, instead of $30km/s$, Einstein's theory predicted an expected result of $0km/s$. At first glance, one might not view $8km/s$ as confirmation of $0km/s$. However many scientists, especially those who support relativity theory, describe Michelson and Morley's $8km/s$ result as experimental error. They argue that the Michelson–Morley experiment actually produces $0km/s$ and that the experimental error of $8km/s$ can be ignored. This interpretation of $0km/s$ as the experiment's result is viewed as confirmation of Einstein's theory.

The Michelson–Morley experiment is one of the best–known, best–reviewed experiments associated with relativity theory. Because it has been thoroughly reviewed for more than 120 years, we immediately ask the question: *If something is wrong with the Michelson–Morley experiment, why wasn't it found earlier?* To answer this question, we have to explain how the Michelson–Morley experiment works, how they calculated their expected result, and how their experimental device – called an interferometer – works. We will discuss specific mistakes in their equation, which when corrected reveal that the experiment actually worked. Said another way, we will show that the experiment supports a partially dragged aether model.

How the Interferometer Experiment Works

The Michelson–Morley interferometer experiment is based on the idea that if light travels as a wave, the wavelength of that ray of light from the perspective of an observer on Earth appears to increase if it travels in the same direction as the Earth, and appears to decrease if it travels in the opposite direction.

The reason the Michelson–Morley test was so ingenious is that they did not have the benefit of modern–day measuring devices like oscilloscopes, precise stopwatches, or frequency counters. In addition, to conduct their experiment they needed to make their observations while accounting for things they could not control or change, such as the velocity of the Earth. To overcome these obstacles, they created a device – called an interferometer – that was mounted on a special table floating on mercury that could rotate 360 degrees around one axis. This construction minimized distortions caused by passing vehicles that could affect the accuracy of the device. Illustrated in Figure 7–1, the interferometer consisted of several components: a light source, a splitter that would split a ray of light into two beams, several mirrors to reflect light back toward the light source, and a microscope–like device to view the two beams of light as they interfered with each other.

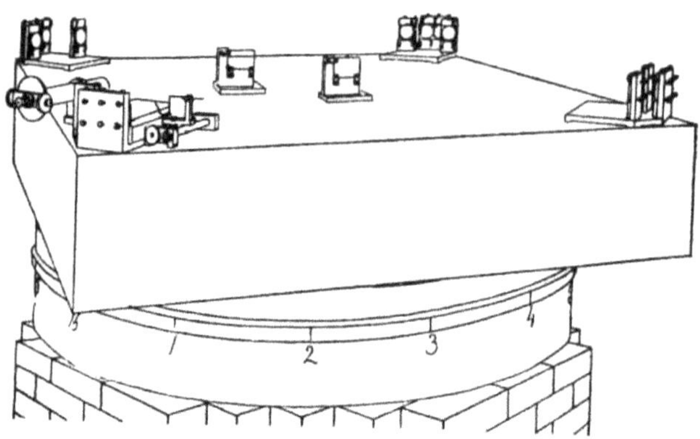

Figure 7–1 The Michelson–Morley interferometer. Source: *On the Relative Motion of the Earth and the Luminiferous Ether*

The operation of the experiment is straightforward. The interferometer emits light from the source, which is immediately split into two beams. The light travels along two perpendicular paths, which they called arms. At this point in the experiment, they know that the two beams are 100% in phase with one another. At the origin where the arms meet are the light source and the measurement device. At both ends of both arms are mirrors. Each beam is sent along perpendicular paths of equal length where they are reflected by the mirrors at the end of each arm. In one case, the ray is sent from the origin to the mirror on the y axis of the inner system where it is reflected back to the origin. In the other case, the ray is sent from the origin to the mirror on the x axis where it is reflected back to the origin. After performing one measurement, the device is rotated 22.5 degrees and a new measurement is observed. Measurements are performed in 22.5 degree increments until the entire device has completed a 360 degree rotation.

A ray of light traveling between the origin and the mirror along the x axis behaves according to the x axis intercept equations. Similarly, a ray of light traveling between the origin and the mirror along the y axis behaves according to the y axis intercept equations. Because the rays travel different distances while the inner system is in motion due to the motion of the Earth around the Sun, the reflected rays should not arrive at the measuring device simultaneously. In other words, while they were sent at the same time, the rays are expected to return to the origin at different times.

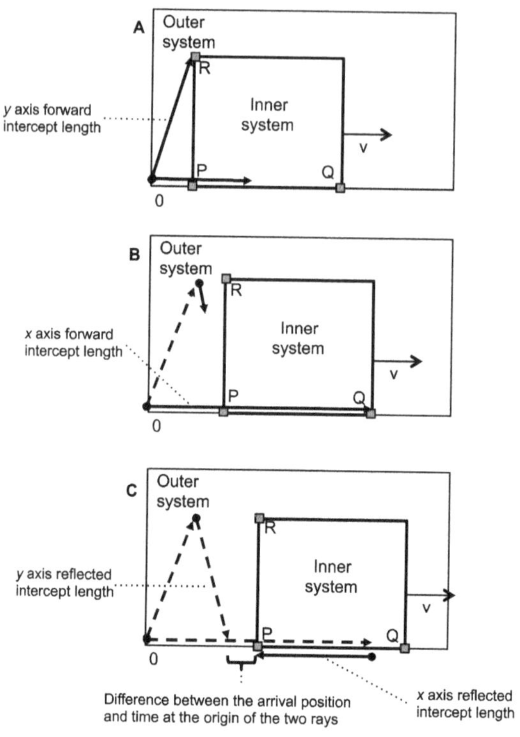

Figure 7–2 Expected behavior of the light rays as they travel from the light source at point P to the mirrors at perpendicular ends – R along the y axis and Q along the x axis – of the interferometer. If the rays move according to the mathematics of a non-nested relationship (aka, a partially dragged model), they will arrive at the origin at different positions. Arrival at different positions means they arrived at different times that can be detected and measured. Note: The length of the arrows in the diagram are not to scale.

As illustrated in Figure 7–2, a beam of light is split at the origin and the two rays are sent along perpendicular paths until they reach mirrors where they are reflected back to the origin (or measuring device) of the moving system. If light travels in an electromagnetic medium then the distance each beam travels before returning to the origin will differ because the device is moving relative to the sun. Because the rays will travel different lengths, they should arrive at the origin at different times. Michelson and Morley expected to measure this difference and use it to compute the Earth's orbital velocity.

Specifically, in Figure 7–2(A) when the first ray travels a distance of n meters with respect to the outer system, it has reached the mirror at R on the y axis; while the second ray, which also travels a distance of n meters, is still in transit to the mirror at Q on the x axis. In Figure 7–2(B), the ray along the x axis has reached the mirror at Q, but the ray on the y axis is well on its way back to the origin. The ray oscillating on the y axis will return to the origin first (not illustrated). In Figure 7–2(C), the second ray arrives at the origin, however the first ray has already arrived. The difference in the arrival time of the two rays is measured and used to calculate the Earth's orbital velocity.

While they did not have a means to measure time or count frequency, they devised a method that used the microscope to measure the movement of the interfering rays of light produced by both returning rays. They conducted measurements at specific intervals as they rotated the device on its axis. If the returning rays of light were still in phase, they would not need to adjust the microscope as the interferometer was rotated. However, if a ray of light travels as a wave through an electromagnetic medium, then the time taken for the ray to oscillate on the x axis would be longer than the time taken for the ray to oscillate on the y axis. The change in distance would cause the interference image, called a fringe, viewed in the microscope to shift to the left or right. They adjusted a screw on the microscope to realign the fringe in the center of the microscope. Each turn of the screw represents a specific size, or fringe shift, that they would use to calculate the Earth's orbital velocity. It is crucial to understand that their experiment measures the number of turns of a screw, not the actually velocity of the Earth. To find the Earth's orbital velocity, they needed to convert the number of screw turns into a velocity using the equation they produced.

Conceptually, the experiment makes sense and their mathematical equations are deceptively easy to produce. The rays of light behave according to the equations developed in Chapter 4. Using Modern Mechanics terminology, Michelson and Morley built their equations using the x axis forward and reflected intercept times, which are:

$$T_{xF} = \frac{D}{c - v} \qquad \text{Eq. 7.6}$$

and

$$T_{xR} = \frac{D}{c + v} \qquad \text{Eq. 7.7}$$

where T_{xF} is the forward intercept time along the x axis, T_{xR} is the reflected intercept time, D is the distance along the x axis, c is the velocity of the ray of light, and v is the Earth's orbital velocity. The meaning of the variable D will be explained shortly.

Similarly, the forward and reflected intercept times for the ray traveling along the y axis are:

$$T_{yF} = T_{yR} = \frac{D}{\sqrt{c^2 - v^2}} \qquad \text{Eq. 7.8}$$

where T_{yF} is the forward intercept time along the y axis, T_{yR} is the reflected intercept time, D is the distance along the y axis, c is the velocity of the ray of light, and v is the Earth's orbital velocity.

The total time for one oscillation along the x axis is:

$$T_x = T_{xF} + T_{xR} \qquad \text{Eq. 7.9}$$

and total time for one oscillation along the y axis is:

$$T_y = T_{yF} + T_{yR} \quad \text{Eq. 7.10}$$

Michelson and Morley reason that if light moves as a wave through an aether, then the difference in the transit time for both rays of light is:

$$\Delta T = T_x - T_y \quad \text{Eq. 7.11}$$

Because this equation can effectively measure the difference in transit time for both rays, it can be used to find the Earth's orbital velocity. Surprisingly, this equation is conceptually and mathematically correct, but it will not produce the correct results when used with their interferometer. In fact, even if we knew with 100% certainty that the Earth traveled through a luminiferous aether at $30km/s$, their experiment would be incapable of measuring this. As you will soon read, their primary mistake is the mistreatment of compound types (eg, frequency and wavelength) as if they were discrete types (eg, time and length). This mistreatment caused them to produce an incorrect expected result equation.

If the interferometer were able to detect and measure time and length, their equation would perform as expected. However, the device does not measure time or length. Remember, they did not use a stopwatch or frequency counter. What they were able to measure was the change in the position of a fringe. In modern terminology, they measure the change in the position of the interference pattern; or simply: the relative change in wavelength caused by the velocity of the Earth as it travels through an aether. To understand the necessary adjustments to the equation, we must examine how a ray behaves when traveling along each arm and what constitutes a fringe.

We will begin with a simplified example, one where the inner system is not in motion. Imagine a stationary light source located at the origin that is emitting a ray of white light. The ray of light has a specific wavelength, λ. The ray travels at velocity c from the source to a stationary mirror. The distance between the source and the mirror is D meters with respect to the interferometer and the time taken to travel the distance is T seconds. Upon reaching the mirror, the ray is reflected back and returns at velocity c to the origin. Illustrated in Figure 7–3, this round–trip journey represents one oscillation. The total distance traveled by the ray is $2D$ meters and the total transit time is $2T$ seconds. This analysis, based on the addition of lengths and times, is consistent with the equations Michelson and Morley developed. The wavelength of the ray of light remains unchanged, because the inner system is not in motion.

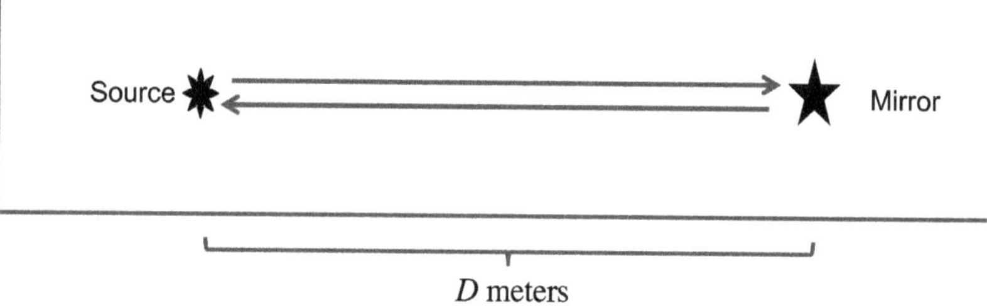

Figure 7–3 A ray of light is emitted from a stationary source toward a stationary mirror where it is reflected back to the source. Because the distance from the source to the mirror is D meters, the total distance the ray travels in one oscillation is $2D$ meters. Since the time taken for the ray to travel from the source to the reflector is T seconds, the total time required to complete one oscillation is $2T$ seconds.

We now ask two critical questions about the ray of light:

1. What is the *length* that best describes the ray of light's movement after one oscillation?

2. What is the *wavelength* that best describes the ray of light after one oscillation?

Both questions can be answered intuitively. Since the distance between the source and the mirror is D meters, the distance of one oscillation is $2D$ meters. For example, if the distance D between the source and the mirror were $300{,}000{,}000m$ and required 1 second to complete, then the total round–trip distance would be $600{,}000{,}000m$ and would require 2 seconds to complete.

The second question is answered by describing the wavelength following one oscillation. We intuitively know that the wavelength of the ray of light did not change, because the inner system is not in motion. Notice that length changes, while wavelength does not. This is an important distinction between these types. While we can add lengths when considering the back–and–forth motion of a ray of light, we cannot sum their corresponding wavelengths.

Figure 7–4 illustrates the association between wavelength and frequency for an oscillating ray of light. A ray of light is emitted from a stationary source to a stationary mirror where it is reflected and travels back to the source. Since the ray is emitted at a specific wavelength and there is a one–to–one relationship between frequency and wavelength, we can count the number of cycles between the source and the mirror as f cycles. The total number of cycles for a complete oscillation is $2f$ cycles. However, since we know the wavelength has not changed, we also know the frequency has not changed. In other words, since the observed wavelength remains λ, the observed frequency must also remain fHz. We can sum cycles, but not cycles per second.

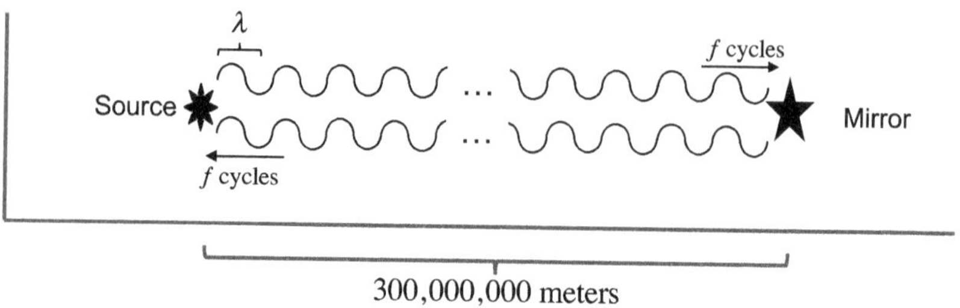

Figure 7–4 The total number of cycles that a ray travels during one oscillation is $2f$, but frequency at the completion of one oscillation remains f cycles per second.

While it is true that the ray will conceptually produce $2f$ cycles, this total is produced over 2 seconds. It is important to distinguish between cycles as a discrete type and cycles per second as a compound type. Frequency is a compound type that is expressed in terms of *cycles per second*, which in this case is:

$2f$ cycles per 2 seconds

Because this answer is expressed as *per 2 seconds*, it is considered a non–standard answer, one that must be expressed in terms of cycles *per second*. Thus, the answer is adjusted and expressed as:

f cycles per second

Not only is this the accepted form for expressing a frequency, it demonstrates that $2f$ is incorrect when referring to the expected frequency following one oscillation. The correct answer is fHz. In fact, if light had a frequency of $2fHz$ we would be unable to complete the experiment because the returning rays of light would be outside the range that the human eye can see! In modern terminology, we use Hertz, or Hz, to represent frequency rather than saying cycles per second, although the two represent the same thing. Michelson and Morley's terminology predates the modern standard, Hertz, which was named after Heinrich Hertz in 1930 by the International Electrotechnical Commission.

We now examine the case where the source and receiver are on a moving inner system. Figure 7–5 illustrates the expected results of the experiment in terms of wavelength. In Figure 7–5(A) the forward and reflected intercepts are added to one other, placing the experiment's expected result much farther to the right. Both the reflected and forward intercepts are expressed in absolute terms. The original wavelength is ignored and the expected result is greater than either individual intercept. As indicated earlier, this answer is mathematically wrong because the forward and reflected wavelengths cannot be summed to produce the expected result. Because the experiment is based on wavelength, summing the intercepts will produce a value that is twice as large as what the experiment will actually measure. Yet, this additive answer is what Michelson and Morley use.

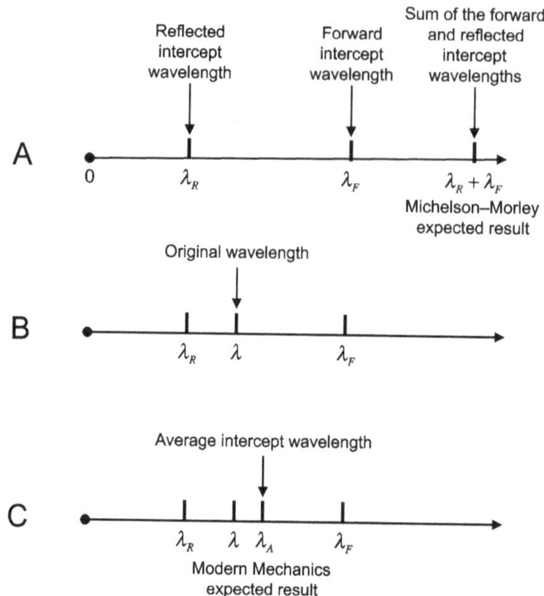

Figure 7–5 In (A), the Michelson–Morley expected result is based on the sum of the forward and reflected intercept wavelengths. The sum is greater than either intercept wavelength and the original wavelength is ignored. In (B), the reflected and forward intercept wavelengths do not shift in the same direction relative to the original wavelength. The reflected intercept wavelength is represented as a shift to the left of the original wavelength and the forward intercept wavelength is represented as a shift to the right of the original wavelength. In (C), the expected result is the average intercept wavelength.

Figure 7–5(B) illustrates the results measured relative to the original wavelength. In this case we would expect one of the shifts to have a higher wavelength than the original ray of light and the other to have a lower wavelength. The reflected intercept wavelength produces a shift to the left of the original wavelength and the forward intercept wavelength produces a shift to the right of the original wavelength. It is very important to recognize that, for the ray traveling along the x axis, we expect to see a visible shift in one direction or the other relative to the original wavelength.

Notice that the forward intercept wavelength is shifted to the right of the original wavelength and the reflected intercept wavelength is shifted to the left. What should we expect as the expected wavelength when the two are combined? Mathematically, use of the sum or difference does not make sense. Figure 7–5(C) illustrates the expected result – the average intercept wavelength – which falls between the reflected and forward intercept wavelengths.

The importance of the proper treatment of types cannot be overemphasized. Michelson and Morley developed equations that are correct and appropriate for discrete types, like time and distance. However, their interferometer is incapable of capturing time and distance measurements. It measures the relative change in wavelength, requiring equations appropriate for the compound types of wavelength and frequency. Specifically, it requires the use of the average intercept wavelength instead of the sum of the forward and reflected intercepts. The average intercept wavelength and frequency equations must be used to combine wavelengths and frequencies.

We are now able to explain the meaning of the D variable. Michelson and Morley believe that their equations work on length, which is confirmed by statements like: "D = distance" in their paper. Notice that the variable D is defined as a length, not a wavelength. Yet, later they use D to represent frequency. This occurs when they say D is "waves of yellow light." In modern terminology, D represents cycles per second of yellow light, or simply as yellow light at DHz. It is crucial to recognize the change in the type of the variable used in their expected result equation and the type that could actually be measured by their interferometer.

A second problem with the Michelson and Morley experiment is that they do not properly adjust their measurements to account for the width of a fringe. They assume that the width of a fringe is one wavelength. One cycle, or wavelength, of a ray of light is illustrated in Figure 7–6(A). Conceptually, we will not observe an interference pattern with one ray alone. Interference patterns require at least two interacting waves. Figure 7–6(B) conceptually illustrates the formation of two peaks and one valley resulting from the interaction to two rays. Two waves produce two fringes over the span of one wavelength. In other words, each fringe has a width of half a wavelength.

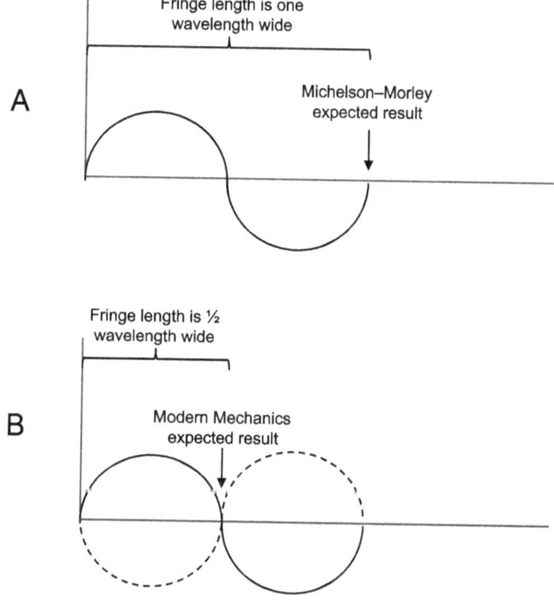

Figure 7–6 In (A), Michelson and Morley define the width of the fringe as the length of one wavelength. In (B), two interacting waves will produce two fringes over the span of one wavelength, which means that the width of each fringe is half a wavelength.

This understanding of the width of a fringe represents the second adjustment, which defines the width of a fringe as $\frac{1}{2}\lambda$ rather than λ. This correction is made to their displacement equation and is illustrated in Figure 7–7.

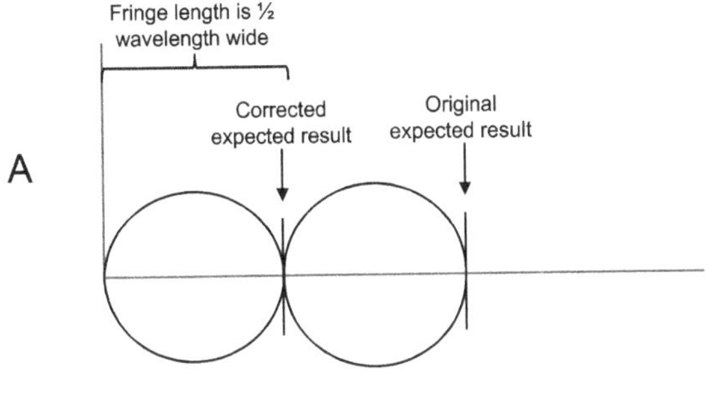

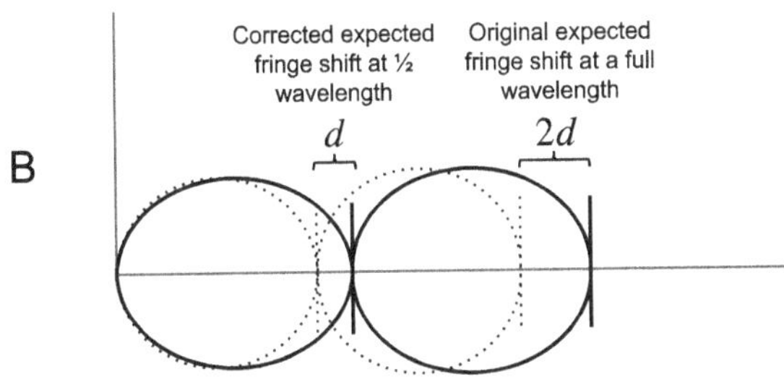

Figure 7–7 Two fringes are produced in the width of one wavelength. In the original Michelson–Morley analysis the fringe calculation is for a wavelength, representing the displacement, $2d$, of two fringes. In the corrected equation, the fringe calculation is performed using half a wavelength, which represents the displacement, d, of one fringe.

A third correction is associated with the behavior of the device and how the measurements are captured and used. We cannot know with certainty whether the microscope will need to be adjusted to the left or right to realign the image in the center. Furthermore, just because our last turn of the screw was in a certain direction, that doesn't mean that the next turn of the screw will be in the same direction. This means we must account for every turn of the screw, regardless of the direction of the turn.

Associated with the counting of every turn of the screw, we also align the expected result and actual result measurements, using 22.5 degree intervals throughout, instead of an expected result that was defined at 90 degree intervals. This correction will facilitate our ability to compare the Michelson–Morley results with the revised equations.

Comparison of the Original and Revised Results

We now compare the Michelson–Morley equation against the corrected equation. Table 7–1 summarizes the results when their data is evaluated using the original Michelson–Morley equation. The preceding discussion notwithstanding, they computed an Earth orbital velocity of $8km/s$. Statistically, we are 99.9% sure that the Earth's orbital velocity falls between $6km/s$ and $10km/s$. This result means that the experiment does not support any theory that expects $30km/s$ or $0km/s$. As stated in Chapter 1, their results failed to support any aether drag model. It also fails to support relativity theory.

Measurement	Actual Results			
	Average micrometer divisions per 360 degree rotation of the Interferometer	Number of micrometer divisions per 22.5 degree rotation of the Interferometer	Displacement per 22.5 degree rotation of the Interferometer (Hertz)	Computed Earth Velocity (approx. - meters per second)
Morning Results				
Jul 08	31.00	1.9375	0.0388	9340
Jul 09	22.60	1.4125	0.0283	7975
Jul 11	22.20	1.3875	0.0278	7905
Morning Average	25.27	1.5792	0.0316	8425
Evening Results				
Jul 08	21.10	1.3188	0.0264	7990
Jul 09	19.40	1.2125	0.0243	7390
Jul 12	22.20	1.3875	0.0278	7905
Evening Average	20.90	1.3063	0.0261	7661
Overall Average	23.08	1.4427	0.0289	8060
Standard Deviation				655

Table 7–1 Michelson–Morley results when the data is evaluated using the original Michelson–Morley equation. They computed an Earth orbital velocity of $8km/s$.

Table 7–2 summarizes the results of the Michelson–Morley experiment when the data is analyzed with the corrections discussed above applied. The proper analysis reveals that the experiment detected an Earth orbital velocity of $32km/s$, which is extremely close to the expected result of $30km/s$. In fact, given the level of accuracy of their device, this could be considered a bull's–eye.

Measurement	Actual Results			
	Average micrometer divisions per 360 degree rotation of the Interferometer	**Number of micrometer divisions per 90 degree rotation of the Interferometer**	**Displacement per 90 degree rotation of the Interferometer (Hertz)**	**Computed Earth Velocity (approx. - meters per second)**
Morning Results				
Jul 08	31.00	7.7500	0.1550	37325
Jul 09	22.60	5.6500	0.1130	31870
Jul 11	22.20	5.5500	0.1110	31590
Morning Average	25.27	6.3167	0.1263	33700
Evening Results				
Jul 08	21.10	5.2750	0.1055	30800
Jul 09	19.40	4.8500	0.0970	29530
Jul 12	22.20	5.5500	0.1110	31590
Evening Average	20.90	5.2250	0.1045	30500
Overall Average	23.08	5.7708	0.1154	32210
Standard Deviation				2689

Table 7–2 Michelson–Morley results when the data is evaluated using the corrected Modern Mechanics–based equation. The revised equation computes an Earth orbital velocity of $32km/s$.

The net result of these changes is that the corrected equation properly converts the raw data, or number of screw turns, into an Earth orbital velocity of $32km/s$. The Michelson–Morley experiment produced a computed Earth orbital velocity that is remarkably close to the expected result of $30km/s$. In other words, the Michelson–Morley experiment was a success, providing support for a partially dragged aether model.

Common Objections

One criticism of this analysis suggests that the adjustments simply manipulate the data to obtain the desired result. This criticism ignores the mathematical and conceptual rationale behind the corrections. The response to this criticism is to show that the same correction can be applied to another experiment to produce the same expected result.

In 1933, Dayton Miller performed a repeat interferometer experiment. His work is based on the idea that the failure of the Michelson–Morley experiment was due to an insufficient level of accuracy in their interferometer. He built a new interferometer; one that he said would perform to a higher degree of accuracy than the Michelson–Morley device. Similar to the Michelson–Morley experiment, Miller collects and analyzes the data, but does not reach the expected result of $30km/s$. He computes an Earth orbital velocity of approximately $11km/s$. While his result was larger than the Michelson–Morley result, like the original, it does not statistically support any theoretical explanation for a luminiferous aether. Not only does his experiment fail to support any aether–based theory, his answer of $11km/s$ did not confirm the Michelson–Morley result of $8km/s$. Being further from $0km/s$, his result cannot be explained away as experimental error. Many scientists have attempted to discredit Miller's work, because it does not support relativity theory.

Fortunately, like Michelson and Morley, Miller provides his measurements, enabling us to determine how the corrected equations perform using his data. An analysis using his data and the revised equation reveals that Miller's experiment found an Earth orbital velocity of $30km/s$, with an error of $0.3km/s$! Not only does he accurately measure the Earth's velocity, he met his goal of producing a device that improves upon the accuracy of the first device.

This result demonstrates that the corrected equations accurately evaluate the data sets of multiple experiments. In addition, the corrected Miller result provides confirmation of the Michelson–Morley experiment, since both produce $30km/s$.

A second objection suggests that the expected result should be greater than $320km/s$ instead of $30km/s$. This answer is based on experiments associated with the Cosmic Background Explorer (COBE). This argument is based on the belief that a COBE–based theory is correct, one that has some other frame of reference than the Sun. The response to this objection is to acknowledge that if the corrections to the algorithm are right, then we must accept the new results, regardless of our belief in any specific aether–based hypothesis. Additionally, since COBE and the interferometer were designed to measure different phenomena, it is possible that the two experiments are not mutually exclusive.

Michelson–Morley Summary

The Michelson–Morley experiment is one of the most important experiments associated with the electromagnetic force and modern physics. While many scientists supported the idea of a luminiferous aether, the reason this idea did not prevail is due to the failed result from the Michelson–Morley experiment. The experiment's failed result meant that scientists had to look for an alternative to the aether drag model. Einstein's theory of relativity would later fill the void. However, as discussed, relativity theory is invalid and the Michelson–Morley experiment was improperly analyzed.

Fortunately, we have explained how to properly interpret the Michelson–Morley data, revealing the correct result of $30km/s$. Their experiment was a success. This finding alone has the potential to rewrite physics because it supports an electromagnetic aether that behaves according to the rules of geometric transformations. In fact, had the Michelson–Morley experimental data been properly analyzed from the start, it would have supported the partially dragged hypothesis and eliminated the need for an alternative explanation. This conclusion is extremely exciting and opens the door for new and revised research in the luminiferous aether.

Energy and the Ives–Stilwell Experiment

The year 1905 was an amazing year for Einstein. In less than twelve months, he published four papers that would forever change the scientific landscape. Although his papers addressed different characteristics of the electromagnetic force, collectively they form the foundation of relativity theory. Many people have never read Einstein's paper, *On the Electrodynamics of Moving Bodies*, yet those same people have heard of relativity, the theory it introduced. While relativity theory is Einstein's most well–known theory, it was left to another paper, *Does the Inertia of a Body Depend on its Energy Content*, to introduce his most famous equation:

$$E = mc^2 \qquad \text{Eq. 7.12}$$

Arguably, Equation 7.12 is Einstein's most well–recognized equation. Even if people do not know where the equation originates or what it means, they can often recite it without much effort. To understand the importance of this energy equation, we must examine how it was derived and experimentally confirmed. However, before we can examine Einstein's derivation, we must explore an important mathematical technique called a Taylor series.

It is hard to imagine a time without computers or the Internet: a time when a math equation could take hours, days, or weeks to precisely solve, if it could be solved at all. Fortunately, mathematicians understood that many equations did not require precise solutions to be useful. They realized that many difficult to precisely solve equations were actually quite easy to approximate to a high degree of accuracy. A common tool used to produce approximate equations is called a Taylor series, which is an infinite series of expressions that approximates a precise equation. Said simply, a Taylor series allows us to easily approximate an answer that otherwise might be difficult or impossible to solve precisely.

Approximations are not the same as precise answers. In fact, approximations must be discussed in terms of how close they are to precise answers, using a measurement called the **degree of accuracy**. The discipline associated with degrees of accuracy, also referred to as levels of accuracy, is so important that it is addressed in specialized courses like numerical analysis, which is often taught as part of computer science, engineering, and mathematics.

To illustrate the relationship between precise answers, approximations, and degrees of accuracy, consider the value π, which begins:

$$\pi = 3.14\,\underline{1}59\,265\,359\,\ldots \qquad \text{Eq. 7.13}$$

and the value $\frac{22}{7}$, which begins:

$$\frac{22}{7} = 3.14\underline{2}85\,714\,286\,... \qquad \text{Eq. 7.14}$$

Equations 7.13 and 7.14 are not the same. They differ in their third significant digit, which is shown with an underline. We cannot use the transitive relation with equations 7.13 and 7.14 to conclude that π is equal to $\frac{22}{7}$. In fact the opposite is true; we can easily show that:

$$\pi \neq \frac{22}{7} \qquad \text{Eq. 7.15}$$

Now we will use equations 7.13 and 7.14 to conceptually demonstrate how a Taylor series works, as well as highlight an all–too–common mistake that occurs when approximations are mistreated as precise values. Begin by rewriting π as the series:

$$\pi = 3.14 + 0.00159 + 0.00000265 + 0.00000000359 + ... \qquad \text{Eq. 7.16}$$

This equation is constructed in a specific way, so that the accuracy improves as more expressions are added to the right of the equation. It should come as no surprise that when you add each expression given in Equation 7.16, you would arrive at the same value given by Equation 7.13.

While we can increase the accuracy of the series by adding more expressions to the equation, we must recognize two important considerations. First, while the inclusion of a new expression adds to the equation's accuracy, it also increases the amount of time and complexity required to manually solve. Second, we can improve the accuracy of the equation to a point where it exceeds the accuracy of that which can be measured or observed. For example, there is no reason to calculate the expected speed of a marathon runner in milliseconds if your stopwatch is only capable of measuring to the nearest tenth of a second.

Often introduced as a useful tool in calculus, Taylor series are approximations, albeit good ones. They enable us to use part of the series to easily arrive at answers that are "accurate enough" for a particular purpose. For example, ignoring all but the first expression of the series given in Equation 7.16, we would say:

$$\pi = 3.14 \qquad \text{Eq. 7.17}$$

Many students are introduced to Equation 7.17 in elementary or high–school math class. In fact, you may have learned and memorized this useful equation. So, it may surprise you to learn that it is incorrect. The use of the equals sign suggests that π equals 3.14 precisely. Said another way, it means that π = 3.140. This level of precision requires that the third significant digit of π is zero. It is true that in cases where accuracy to two significant digits is all that is required, this equation will be sufficiently accurate. However, when more accuracy is required we would need to include more expressions from the series given in Equation 7.16.

Since we know from Equation 7.13 that the third significant digit of π is one, we easily conclude that Equation 7.17 is an approximation of π, not its precise value. As an approximation, it is properly written using the approximation operator as:

$$\pi \approx 3.14 \qquad \text{Eq. 7.18}$$

Notice the difference between the "=" operator that is used to express precision and the "≈" operator that it used to express approximation. In many cases the equals and approximation operators can be used interchangeably without adversely affecting the result, but in other cases this misuse will produce difficult–to–detect math mistakes. As an example, consider the second value introduced above, $\frac{22}{7}$, which when expressed as a series is:

$$\frac{22}{7} = 3.14 + 0.00285 + 0.00000714 + 0.00000000286 + \ldots$$

Eq. 7.19

Similar to the preceding discussion involving π, when all but the first expression in Equation 7.19 are ignored, this equation is written as:

$$\frac{22}{7} = 3.14 \qquad \text{Eq. 7.20}$$

Like Equation 7.17, Equation 7.20 is incorrect because it is written as a precise statement when it is actually an approximation. When written correctly using the approximation operator, it is:

$$\frac{22}{7} \approx 3.14 \qquad \text{Eq. 7.21}$$

Because equations 7.17 and 7.20 are mistreated as precise statements, not as approximations, they are improperly used with the transitive relation to produce:

$$\pi = 3.14 = \frac{22}{7} \qquad \text{Eq. 7.22}$$

which becomes:

$$\pi = \frac{22}{7} \qquad \text{Eq. 7.23}$$

Equation 7.23, which shows that π is equal to $\frac{22}{7}$, is a direct contradiction of Equation 7.15, which shows that they are not equal. Both equations cannot be true. Understanding the root cause of these conflicting equations and why Equation 7.23 is incorrect requires us to distinguish between the approximation and equals operators. In this case, equations 7.22 and 7.23 are both incorrect, because the transitive relation cannot be used with approximations.

Consider equations 7.18 and 7.21, which are properly written as approximations. While it is true that:

$$\pi \approx 3.14 \approx \frac{22}{7} \qquad \text{Eq. 7.24}$$

we cannot use the transitive relation on the approximation operator to conclude that π is approximately $\frac{22}{7}$ without also considering the degree of accuracy. To do otherwise would lead to a myriad of mathematical problems, such as concluding that $\pi = e$, since both are approximately 3 to the nearest integer. In reality, we would never say that π, which begins as 3.141, is equal to e, which begins as 2.718. Yet, we can say that both are approximately 3 to the nearest integer.

While it is easy to show that:

$$\pi \approx \frac{22}{7} \qquad \text{Eq. 7.25}$$

to a certain degree of accuracy, this conclusion is based on other mathematical techniques. Specifically, it involves assessing the relative closeness of the values prior to adjusting the equations using rounding, truncating, or another technique.

Approximations maintain their validity for specific degrees of accuracy. Beyond this level of accuracy, the approximations are assumed to be inaccurate. Notice that when the required level of accuracy is two significant digits, π and $\frac{22}{7}$ can be used interchangeably, since both begin as 3.14 and are accurate to two significant digits. However, when the desired level of accuracy exceeds two significant digits, they can no longer be used interchangeably and their results will diverge.

Returning to Einstein's well–known energy equation: you will find that $E = mc^2$ does not clearly appear in his paper. Instead it is written textually as the phrase "mass diminishes by $\frac{L}{c^2}$." To see how this phrase is translated into his energy equation, we have to consider the last step in his derivation. Begin by replacing the variable L with the variable E to align Einstein's nomenclature with modern convention. Second, replace Einstein's phrase "mass diminishes by" with "$m =$ " to produce:

$$m = \frac{E}{c^2} \qquad \text{Eq. 7.26}$$

The simple rearrangement of terms in Equation 7.26 yields the $E = mc^2$ equation, which is well–known and extremely useful. However, in light of our earlier discussion of the validity of Einstein's relativity theory we must ask a very important question: *Is it correct?* To answer this question, we must examine Einstein's entire derivation.

Einstein's derivation of the energy equation is straightforward. He begins by assuming that relativity theory is valid; an assumption we have already shown is incorrect. However, for the purpose of this analysis, let's take Einstein's assumption as true; that relativity theory is valid. Einstein then performs several mathematical steps to arrive at:

$$K_0 - K_1 = L\left(\frac{1}{\sqrt{1-\frac{v^2}{c^2}}} - 1\right) \qquad \text{Eq. 7.27}$$

For the purpose of this discussion, assume the mathematics Einstein follows to arrive at Equation 7.27 is sound. This is an important equation, one that was previously introduced in Chapter 6. Although Einstein uses the variable L instead of x', which is used in the Modern Mechanics derivation, both equations are equivalent.

Equation 7.27 defines the difference between an initial value and the scaled value resulting from using Einstein's x axis transformation. Without the aid of modern tools like a calculator or a computer, this equation would be difficult to precisely solve due to the radical in the denominator. Fortunately, this equation can be approximated using a Taylor series. The Taylor series approximation for Equation 7.27 is:

$$K_0 - K_1 = \frac{1}{2}\frac{L}{c^2}v^2 + \frac{3Lv^4}{8c^4} + \frac{5Lv^6}{16c^6} + \frac{35Lv^8}{128c^8} + \ldots \qquad \text{Eq. 7.28}$$

As discussed earlier, each expression that is added to the right of the series given in Equation 7.28 improves the accuracy of the result. Similar to the earlier discussion on approximation, Einstein ignores all but the first expression of the Taylor series to conclude:

$$K_0 - K_1 = \frac{1}{2}\frac{L}{c^2}v^2 \qquad \text{Eq. 7.29}$$

Without the aid of computing technology, Equation 7.29 is far easier to manually solve than Equation 7.27. As previously discussed, Equation 7.29 is incorrect because the use of the equals sign defines a level of precision that is not present. When properly stated using the approximation operator, Equation 7.29 is written as:

$$K_0 - K_1 \approx \frac{1}{2}\frac{L}{c^2}v^2 \qquad \text{Eq. 7.30}$$

At this point Einstein's states that the relationship between Equation 7.29 and the energy equation $E = mc^2$ is obvious. However, it is only obvious to those who know that the kinetic energy equation is:

$$K = \frac{1}{2}mv^2 \qquad \text{Eq. 7.31}$$

Einstein further assumes that the expressions on the left–hand side of equations 7.29 and 7.31 represent the same thing; that $K = K_0 - K_1$, and then uses the transitive relation to conclude:

$$\frac{1}{2}mv^2 = \frac{1}{2}\frac{L}{c^2}v^2 \qquad \text{Eq. 7.32}$$

Einstein uses algebraic simplification of canceling like terms to arrive at:

$$m = \frac{L}{c^2} \qquad \text{Eq. 7.33}$$

With the variable substitutions previously discussed to replace L with E, we arrive at Einstein's energy equation $E = mc^2$. However, Einstein's energy equation is incorrect. Because Equation 7.29 is an approximation, it cannot be written with the equals sign. It should be properly written as Equation 7.30, which is an approximation. As an approximation, Equation 7.30 cannot be combined with Equation 7.31 using the transitive relation to arrive at Equation 7.32.

While we cannot use the transitive relation to produce Einstein's energy equation, we can use other techniques to conclude:

$$\frac{1}{2}mv^2 \approx \frac{1}{2}\frac{L}{c^2}v^2 \qquad \text{Eq. 7.34}$$

to a certain degree of accuracy. Canceling like terms, we conclude:

$$m \approx \frac{L}{c^2} \qquad \text{Eq. 7.35}$$

although the degree of accuracy may change. Replacing the variable L with E and rearranging the terms, it follows that:

$$E \approx mc^2 \qquad \text{Eq. 7.36}$$

which is only accurate to a specific degree of accuracy. While $E = mc^2$ is not true because it is written as a statement of precision, $E \approx mc^2$ is true because it is properly written as an approximation. Since Equation 7.36 is an approximation, we must use caution in making definitive statements that associate m with $\frac{L}{c^2}$. The best we can say is $E \approx mc^2$.

The conclusion that Einstein's energy equation is an approximation raises two very interesting questions:

1. Is Einstein' derivation the only way of arriving at $E \approx mc^2$?

2. Does $E \approx mc^2$ still apply or does another equation apply?

To answer these questions, once again begin by examining Einstein's derivation starting with Equation 7.27. As mentioned previously, this equation can be rewritten using Modern Mechanics notation by replacing Einstein's L variable with x' and by replacing the expression on the left–hand side of the equals sign, $K_0 - K_1$, with ΔL_R. To use this equation in analyzing a key experiment, we also replace x' with λ. After these straightforward notational changes are applied, the equation is rewritten as:

$$\Delta L_R = \lambda\left(\frac{1}{\sqrt{1-\frac{v^2}{c^2}}} - 1\right) \qquad \text{Eq. 7.37}$$

which was previously introduced as Equation 6.20. To remain consistent with the discussion on types earlier in this chapter, we assume the proper derivation of this equation for wavelength as a compound type instead of length as a discrete type.

When Equation 7.37 is expressed as a Taylor series using Modern Mechanics notation, it is written as:

$$\Delta L_R = \frac{1}{2}\frac{\lambda}{c^2}v^2 + \frac{3\lambda v^4}{8c^4} + \frac{5\lambda v^6}{16c^6} + \frac{35\lambda v^8}{128c^8} + \ldots \qquad \text{Eq. 7.38}$$

Equation 7.38 is simply a restatement of Equation 7.28, which is the Taylor series for Einstein's Equation 7.27. The use of Modern Mechanics notation will enable us to easily compare the accuracy of Einstein's equation with the Modern Mechanics equation.

As discussed in Chapter 6, Modern Mechanics provides an equation that measures the change in the average intercept length. Originally defined as Equation 6.19, this equation is restated here as:

$$\Delta L_M = x'\left(\frac{1}{1-\frac{v^2}{c^2}} - 1\right) \qquad \text{Eq. 7.39}$$

As discussed in chapters 4 and 6, relativity theory and Modern Mechanics use the x' variable differently. In the relativity equation, the x' variable represents a full wavelength. In Modern

Mechanics, the x' variable represents the length of a segment. In other words, it represents half a wavelength. Substituting $\frac{\lambda}{2}$ for x' in Equation 7.39, the Taylor series is:

$$\Delta L_M = \frac{1}{2}\frac{\lambda}{c^2}v^2 + \frac{1}{2}\frac{\lambda v^4}{c^4} + \frac{1}{2}\frac{\lambda v^6}{c^6} + \ldots \qquad \text{Eq. 7.40}$$

Comparing the Taylor series in Equation 7.40 with that of Equation 7.38, notice that both begin with the same first expression, but they differ in their second and subsequent expressions. In other words, the equations are not equal. Mathematically, we conclude:

$$\Delta L_R \neq \Delta L_M \qquad \text{Eq. 7.41}$$

Similar to our earlier discussion on approximation, if we ignore all but the first expression of equations 7.38 and 7.40, and use the equals operator instead of the approximation operator, we produce:

$$\Delta L_R = \Delta L_M = \frac{1}{2}\frac{\lambda}{c^2}v^2 \qquad \text{Eq. 7.42}$$

The use of the transitive relation simplifies this equation to:

$$\Delta L_R = \Delta L_M \qquad \text{Eq. 7.43}$$

Equation 7.43 is a direct contradiction of Equation 7.41. Both statements cannot be true. Equation 7.43 is incorrect because Equation 7.42 is incorrect. Since all but the first expression of the Taylor series are ignored, Equation 7.42 is incorrectly written using the equals operator. It is properly written using the approximation operator as:

$$\Delta L_R \approx \Delta L_M \approx \frac{1}{2}\frac{\lambda}{c^2}v^2 \qquad \text{Eq. 7.44}$$

Following similar steps as discussed earlier, we combine Equation 7.44 with the kinetic energy equation to arrive at $E \approx mc^2$. This answers the first question: *Einstein's relativity equations are not the only equations that will produce his energy equation.* It also begins to answer the second question, because the energy equation is an approximation and not a precise statement. Later in this section, we will re–examine the relationship between the Modern Mechanics equations and the kinetic energy equation.

Interestingly, Einstein's relativity and energy equations make extremely useful predictions in a number of experiments. One of these foundational experiments is called the Ives–Stilwell experiment. Conducted in 1938 by Herbert Ives and G. R. Stilwell, the experiment was designed to measure the shift in the length of one–half a wavelength and to measure the Doppler displacement of a hydrogen atom in a contained canal ray tube. It used experimental velocities of 0.5% the speed of light. Said simply, the experiment was designed to demonstrate Einstein's time dilation effect predicted by his relativity equations. However, since the analysis of the experiment uses Equation 7.37 (and therefore Equation 7.27), it would also provide experimental support for Einstein's energy equation.

Prior to the author's introduction of the Model of Complete and Incomplete Coordinate Systems, the predecessor to Modern Mechanics, relatively theory was the only idea widely accepted as explaining the Ives–Stilwell experiment. An experiment that is explained via one theory alone creates a risk that the experiment might be considered proof of that single theory. In reality, there could be many theoretical explanations, although perhaps when the experiment was analyzed those theories had not yet been developed. In this case, Modern Mechanics represents a theory that had not yet been developed in 1938 when the experiment was originally conducted and analyzed.

As discussed in Chapter 1, any theory that intends to serve as a unified model must match or exceed the accuracy of any theory it intends to replace. Improved accuracy is essential, since any theory that fails to improve on the predictive capabilities of the historical model is unlikely to unseat the established model. Because Einstein's Equation 7.38 and the Modern Mechanics Equation 7.40 share the same first expression in their respective Taylor series, they will produce results that are indistinguishable from one another to a certain level of accuracy. Beyond this level of accuracy, however, their results will vary. This means that, to a certain degree of accuracy, the Modern Mechanics and relativity equations will provide similar, if not identical, answers. However, when the equations are specified to a greater degree of accuracy or precisely given, they will produce different answers. In the ideal case, Modern Mechanics will provide answers for the Ives–Stilwell experiment that exceed the accuracy of Einstein's equations when increased degrees of accuracy are considered.

The Ives–Stilwell experiment is extremely interesting for three reasons. First, as mentioned earlier, it is thought that relativity theory is the only explanation for its results. Second, it appears to provide experimental confirmation of Einstein's special relativity and energy equations. Third, and most importantly, it provides experimental results at a level of accuracy that can distinguish between the expected results of the Modern Mechanics and relativity equations.

In the analysis that follows, we take advantage of modern computing technology, eliminating the need to express equations 7.37 and 7.39 as Taylor series approximations. However, we will use the Taylor series to explain the difference in the results.

The experiment measured two results: The first was the change in length of one–half a wavelength, referred to by Ives and Stilwell as the "shift in the center of gravity." The second was the measurement of the Doppler shift displacement. Ives and Stilwell computed the displacement as the difference between the approaching and receding observations. We will examine the results for the shift in the center of gravity, which is described by Equation 7.37.

The Ives–Stilwell data was collected and analyzed using the relativistic Equation 7.37 and the Modern Mechanics Equation 7.39. As summarized in Table 7–3, the relativistic equation, also referred to as the relativistic Doppler shift, produces useful results. The observed shift represents the experiment's actual result for each test case, which is represented by the plate number. The relativistic accuracy is the difference between the relativistic Doppler expected result and the observe shift.

	Expected Displacement	Actual Results	Accuracy
Plate	Relativistic Doppler	Observed Shift	Relativistic Doppler
169	10.36	10.35	0.01
160	14.04	14.02	0.02
163	15.42	15.40	0.02
170	16.52	16.49	0.03
165	14.09	14.07	0.02
172	18.71	18.67	0.04
172	15.16	15.14	0.02
177	21.42	21.37	0.05
total	125.72	125.51	0.21
mean	15.72	15.69	0.03
stddev	3.31	3.2971	0.01

Table 7–3 Results of the Ives–Stilwell experiment analyzed with the relativistic equation. The relativistic equation has a total error of 0.21 and an average error of 0.03. The difference between the expected result and the observed shift is explained as experimental error. Note that all values are measured in λ.

Einstein's equation produces useful answers. The amount of error between the expected result and the experimental results averaged 0.03λ. While it is interesting to note that the size of the error increases with the size of the expected result, the error for the entire experiment is sufficiently small and is explained as experimental error. Said another way, the error between the expected and actual results is attributed to weaknesses in the experimental or measurement device, not to a problem in the relativistic equation.

Fortunately, Einstein's equations are not the only ones that explain the Ives–Stilwell experiment. As illustrated in Table 7–4, the Modern Mechanics equation also provides accurate answers and performs better than the relativistic equations.

	Expected Displacement		Actual Results	Accuracy	
Plate	Relativistic Doppler	Modern Mechanics	Observed Shift	Relativistic Doppler	Modern Mechanics
169	10.36	10.35	10.35	0.01	0.00
160	14.04	14.02	14.02	0.02	0.00
163	15.42	15.40	15.40	0.02	0.00
170	16.52	16.49	16.49	0.03	0.00
165	14.09	14.07	14.07	0.02	0.00
172	18.71	18.67	18.67	0.04	0.00
172	15.16	15.14	15.14	0.02	0.00
177	21.42	21.37	21.37	0.05	0.00
total	125.72	125.51	125.51	0.21	0.00
mean	15.72	15.69	15.69	0.03	0.00
stddev	3.31	3.30	3.2971	0.01	0.00

Table 7–4 Results of the Ives–Stilwell experiment analyzed using the relativistic and Modern Mechanics equations. The Modern Mechanics equations precisely match the observations. Note that all values are measured in λ.

The predicted Modern Mechanics results are a precise match for every experimental test. Comparatively speaking, the Modern Mechanics equation performs better than the relativity equations. Notice that there is no measurable error that requires explanation and accuracy does not diminish as the size of the expected result increases. The fact that the Modern Mechanics equation produces exact matches in each instance suggests that the error is associated with Einstein's relativistic equation, not the measuring device.

As discussed earlier, the similarity in answers is due to the same expression appearing as the first in their respective Taylor series. A difference in results cannot be determined using the first expression of the Taylor series alone. Improved accuracy requires the use of the precise equation or the inclusion of more expressions from each series. In tables 7–3 and 7–4, if we round the expected results to one significant digit of accuracy, both models would predict identical results. However, when we include the second significant digit, we can distinguish between the equations and compare their accuracy. Comparing the error, which is the difference between the expected and actual results, the Modern Mechanics equation performs better than Einstein's equation. This satisfies a key requirement for Modern Mechanics to serve as a unified model.

Because the Modern Mechanics equation produces better results for the Ives–Stilwell experiment than the relativistic equation, we must properly derive the kinetic energy equation so that it is consistent with a Modern Mechanics non–nested system relationship. We begin by returning to the Modern Mechanics nomenclature where v is the velocity of the moving system and w is the velocity of the oscillating system. The kinetic energy equation is:

$$K = F\xi \qquad \text{Eq. 7.45}$$

where K is energy, F is force, and ξ is distance. Since $F = ma$ and ξ is the intercept distance given by Function 4.6, Equation 7.45 is rewritten as:

$$K = ma * L(x,y,z,v_x,s,w) \qquad \text{Eq. 7.46}$$

Equation 7.46 is the **generalized kinetic energy equation**, which can be instantiated as the forward, reflected, and average intercept kinetic energy equations. This equation takes into account the motion of the inner system traveling at v and the oscillating system traveling at w.

Considering the specific case of the average intercept length along the x axis, the average kinetic energy equation is:

$$K = ma * L(x,0,0,v,0,w) \quad \text{Eq. 7.47}$$

In Modern Mechanics and classical mechanics the relationship between velocity and acceleration is defined as:

$$w^2 = 2ax \quad \text{Eq. 7.48}$$

where a is acceleration, w is the final velocity of the oscillating system, and x is the segment length of the inner system. Combining equations 7.47 and 7.48 we arrive at:

$$K = \frac{mw^2}{2\left(1 - \frac{v^2}{w^2}\right)} \quad \text{Eq. 7.49}$$

following function instantiation and algebraic substitution. Equation 7.49 is the **average intercept kinetic energy equation** for an oscillating system in a non–nested system relationship. Notice that Equation 7.49 does not place an upper limit on velocity. However, applicability of this equation is limited to cases where $v < w$. If the velocity of the inner system matches or exceeds that of the oscillating system, then this equation does not apply. The velocity of the oscillating system is not constrained

Unlike Einstein's theory where the increase in energy is attributed to increased mass, Modern Mechanics attributes the change in kinetic energy to the increased distance that the oscillating system travels as a result of the inner system's motion.

When the velocity v of the inner system is zero, Equation 7.49 simplifies into the familiar classical mechanics kinetic energy equation:

$$K = \frac{1}{2} m w^2 \qquad \text{Eq. 7.50}$$

where w is the velocity of the oscillating system. Although derived for a three–system model where the inner system is stationary, this special case can be thought of as a two–system model. Said another way, the classical mechanics kinetic energy equation is a special case of the Modern Mechanics average intercept kinetic energy equation where the velocity of the inner system is zero.

In addition to a revised kinetic energy equation, the wavelength–based Modern Mechanics equations could define energy as a function of wavelength, mass, and velocity. Such an equation might align with other areas in physics, most notably, the Planck–Einstein equation, $E = hf$, where E is energy, f is frequency, and h is the Planck constant. This represents an interesting research area, one where the implications of Modern Mechanics will need to be examined and assessed. Clearly, given the improved performance of the Modern Mechanics equations, this will be an interesting research area

Light and the Double–Slit Experiment

One of the most interesting questions in physics during the late 19th century was: *What is the nature of light?* Specifically, scientists wanted to know whether light behaved like a particle, similar to how a ball might be thrown across a field; or like a wave, similar to how a wave caused by a splashing hand in a swimming pool spreads outward in all directions. Supporters of the particle motion of light included Newton, who believes that rays of light, like arrows flung from a bow, travel along specific trajectories unless their courses are changed in transit. This idea is consistent with Newton's first law of motion, which says: *An object in a state of uniform motion will remain in that state of motion unless a force is applied to it.*

Supporters of the wave motion of light include Christiaan Huygens, who believes that light travels in many directions at once from a central emission point. Interestingly, the particle approach explains many, but not all of the observations associated with light. Similarly, the wave approach explains many, but not all of the observations associated with light. Neither approach explains every observation.

A scientist named Thomas Young devised an experiment that he thought would help answer this question about the nature of light. His experiment, called the double–slit experiment, was very ingenious. Conceptually, the Young double–slit experiment is relatively straightforward and easy to explain. The experiment is comprised of an apparatus consisting of several elements, as illustrated in Figure 7–8. On the far left is an *emission source* that produces particles. Consistent with Modern Mechanics terminology, *waves* and *particles* are generically referred to as objects. These objects travel through the *primary field*, or medium, where they eventually meet a *barrier*. The barrier can be completely closed, it can have *one slit*, or it can have *two slits*. A slit is simply a very small opening in the barrier. Any object that passes through one of the slits then moves through the *secondary field* until it reaches the second barrier, which is called the *observation plane*. His experiment consists of sending light through the slits and observing the pattern that forms along an observation plane.

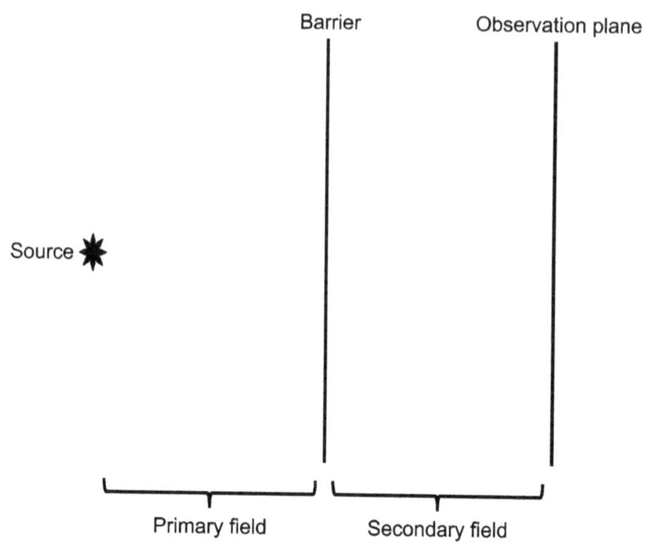

Figure 7–8 Conceptual diagram of the double–slit experiment. The experimental device consists of five distinct areas: the source, the primary field, the barrier, the secondary field, and the observation plane.

A ray of light is emitted by the source. It travels through the primary field until it encounters the barrier. When the barrier does not have any slits, it completely blocks the ray from entering the secondary field. However, when one or more slits are present in the barrier, the ray passes into the secondary field, where it continues until it hits the observation plane.

When light rays are sent through a barrier with one slit, they behave like particles and form a ballistic pattern localized around a point on the observation plane, as illustrated in Figure 7–9. This pattern is consistent with the expected behavior of particles passing through a single slit.

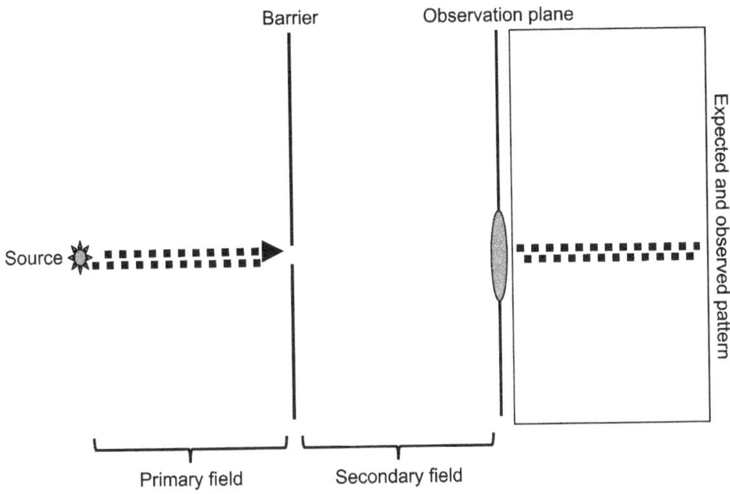

Figure 7–9 Ballistic pattern formed around a single place on the observation plane when light rays pass through the open slit in the barrier. The observed pattern is the same as the expected pattern.

When the light rays pass through a barrier that has two slits, they should behave like particles to form two ballistic patterns, as illustrated in Figure 7–10. Rays that pass through the upper slit should form the ballistic pattern represented by Band 1. Rays that pass through the lower slit should form the ballistic pattern represented by Band 2.

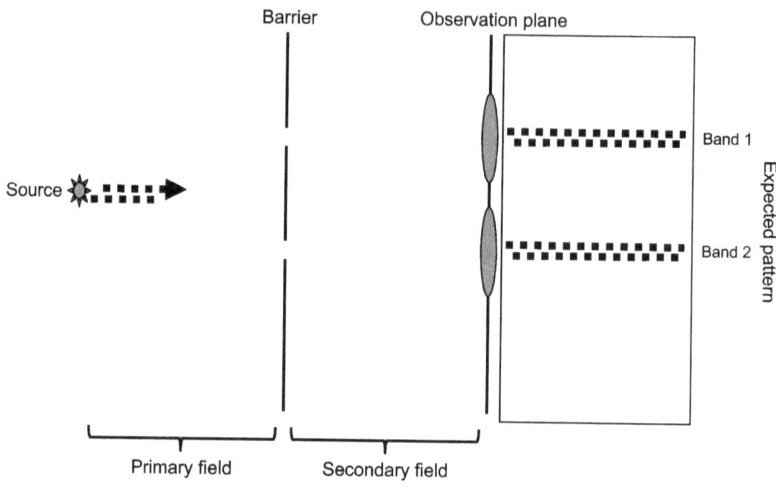

Figure 7–10 The expected pattern at the observation plane when light rays pass through a barrier that has two slits. Light rays with the proper trajectories will pass through one of the two slits, travel through the secondary field, and form one of two ballistic patterns along the observation plane.

Interestingly, Young discovered that light rays did not behave like particles to form the expected ballistic pattern. Instead, they formed an interference pattern on the observation plane, illustrated in Figure 7–11. This pattern aligns with a wave model of light.

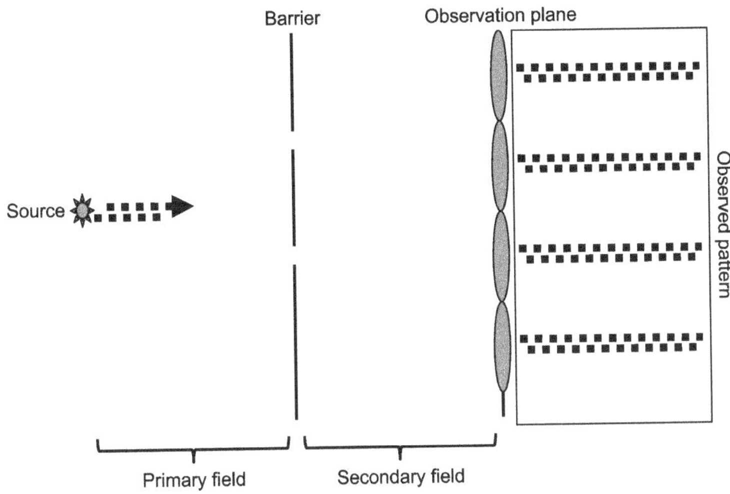

Figure 7–11 The observed interference pattern formed when light rays pass through a barrier with two slits.

The pattern looks similar to the one illustrated in Figure 7–12, which is produced when waves are sent through a barrier with two slits. The interference pattern on the observation plane is similar to the one previously discussed during the analysis of the Michelson–Morley experiment. In this case, the interference pattern is well understood and can be modeled with sound waves and water waves, in addition to light waves.

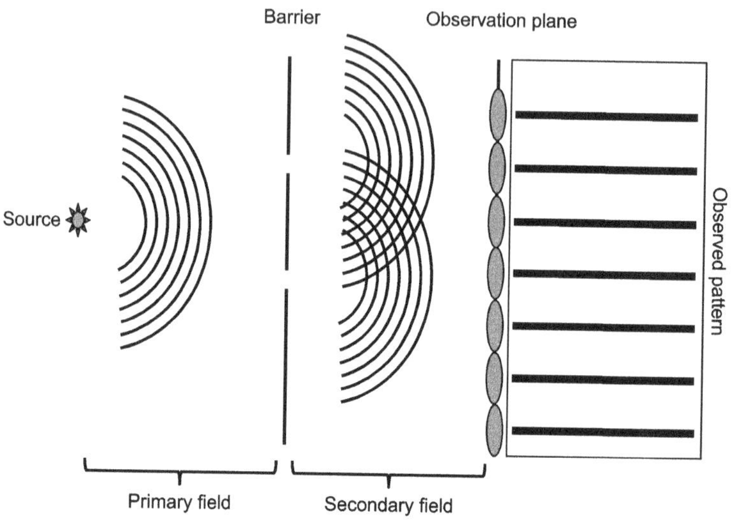

Figure 7–12 Wave behavior through a double-slit barrier. Waves generated by one source travel to the barrier where they pass through two slits. Each slit acts as an emission source in the secondary field. The waves travel through the secondary field where they form an interference pattern on the observation plane.

Technically, calling the pattern illustrated in Figure 7–12 an *interference* pattern is incorrect. These waves are not interfering with each other, but in conceptual terms are simply adding or subtracting their values from the *height* of the other waves in the secondary field. Peaks occur when the waves from both sources arrive at the observation plane at the same time, also called "in phase with one another." Said simply, peaks are formed when one wave is at a peak at the observation plane and the other wave is also at a peak at the same place at the same time. Valleys are formed when both waves arrive at the observation plane 180 degrees out of phase.

An interference pattern is produced through the interactions of more than one wave, each emitted at different times. In Figure 7–13, we have reduced the number of emitted waves to two to show

how interacting waves emitted at different times form peaks and valleys at different locations along the observation plane.

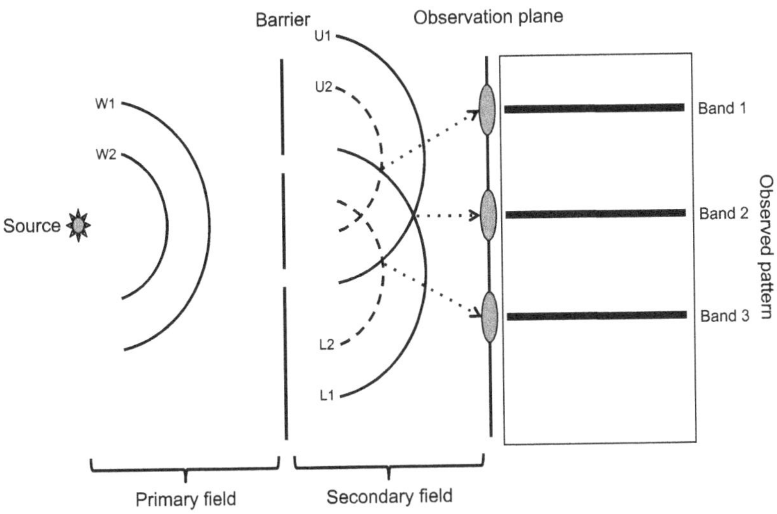

Figure 7–13 Interference patterns are the result of waves emitted by the barrier at different times. First, wave W1 hits the barrier and causes U1 and L1 to be emitted from both slits. Later, wave W2 hits the barrier and causes U2 and L2 to be emitted from both slits. Two waves in the primary field produce four waves in the secondary field, resulting in three interference bands on the observation plane.

Notice that two waves emitted by the source will produce three bands at the observation plane. In the first case, waves emitted from both slits simultaneously will form a peak in the center of the observation plane. For example, when the first wave from the source, represented by W1 reaches the slits in the barrier, each slit will serve as a new emission source for U1 and L1 in the secondary field. When W2 reaches the barrier, each slit will emit U2 and L2 in the secondary field. The crest produced by the interaction of U1 and L1 will form a peak in the center of the observation plane, represented by Band 2. Band 2 is actually formed twice, once when L1 and U1 intersect with the observation plane and again when L2 and U2 intersect with the observation plane. The peaks observed at the center of the observation plane are the result of two waves *arriving at the same time*; one emitted from the upper slit and one from the lower slit, both of which were *emitted at the same time.*

In the second case, when a wave is emitted from the upper slit followed by a wave emitted from the lower slit, then the interference wave will hit a lower location along the observation plane. For example, the first wave U1 is emitted from the upper slit and is followed by a second wave L2 that is emitted from the lower slit. Both waves will arrive at the same time at Band 3 where they will form a peak. Interference waves located lower on the observation plane are due to peaks formed by interacting waves that were *emitted at different times* from the slits in the barrier. While emitted at different times, they *arrive at the same time* at a point below the center of the observation plane.

In the third case, when a wave is emitted from the lower slit followed by a wave emitted from the upper slit, then the interference wave will hit a higher location along the observation plane. For example, the first wave L1 emitted from the lower slit is followed by a second wave U2 emitted from the upper slit. Both waves will arrive at the same time at Band 1, where they will form a peak. Similar to the case just discussed, interference waves located higher on the observation plane are formed by interacting waves that were *emitted at different times* from the slits in the barrier. While emitted at different times, they *arrive at the same time* at a point above the center of the observation plane.

The *interaction* of multiple interference waves – emitted at different times by the slits in the barrier, but arriving at the same time at specific points on the observation plane to form peaks –

explains the formation of the interference pattern. Notice, the visible peak at Band 2 does not occur at the same time as the visible peaks at Bands 1 and 3. While this behavior explains why interacting waves form interference patterns, it does not explain how particles might form a similar interference pattern, as observed in the experiment.

Many scientists believe that the formation of the wavelike pattern of particles as observed in the double–slit experiment cannot be explained using classical mechanics and sought an alternative explanation. Implied in this belief is the idea that this experiment cannot be explained in a way that is consistent with Newton's first law of motion. Early 20th century scientists explored two theories that could explain this experiment: quantum mechanics and pilot wave theory. While both theories were discussed and debated over several decades, by the mid 20th century, quantum mechanics had established itself as the leading theory and pilot wave theory became a historical footnote.

Quantum mechanics offers two alternative explanations for the experiment; one is called *superposition* and the other is called the *multiverse*. While different, they share two important characteristics. First, both suggest that part of a light particle travels through both slits and interacts with another part of the light particle within the secondary field. While the nature of the interaction is unknown, they presume that one of many possible outcomes is observed, causing the interference pattern. Second, both assume the specific state of a light ray cannot be determined until it is measured or observed. In other words, a ray is presumed to exist at every position along the observation plane and the act of measuring or observing it locks it to a specific location.

Regardless of the underlying mechanism, quantum mechanics suggests that light rays behave – on some fundamental level – sometimes like particles and at other times like waves. As you will soon read, this conclusion requires that light rays do things that waves do not. Neither quantum mechanics' explanation takes into account the wave behavior discussed above involving multiple waves emitted at different times to produce the observed pattern. Remember, what is observed as a single pattern along the observation plane is actually the result of many interacting waves emitted at different times from each slit in the barrier.

Quantum mechanics requires that we introduce ideas and behaviors not present with interacting waves. As an example, consider the case of the interference pattern formed along the top of the observation plane. For a particle to behave like a wave to form the pattern, part of the particle would need to go through the lower slit *before* another part of the particle goes through the upper slit. In doing so, the part of the particle going through the lower slit would need to make a decision regarding where on the observation plane it intends to arrive. It would then need to independently change course, as well as retain and communicate this information to the part of the particle that subsequently passes through the upper slit. When the part of the particle later passes through the upper slit, the first part would need to communicate its intended destination, which would need to be received by this second part. Upon receipt, the second part would need to independently alter its course so that it could also arrive at the same location at the same time, since coincident arrival at the same location causes a peak in an interference pattern. This is non–intuitive and violates Newton's first law. Fortunately, another approach – which has previously been discarded – called pilot wave theory, can also explain the double–slit experiment.

Pilot wave theory, as applied to Modern Mechanics, is somewhat easy to visualize. Consider the case of a boat moving at velocity v through a body of water. As it moves, it creates a bow wave that travels outward at velocity w. This motion creates a familiar wake pattern, as well as a bow wave that travels ahead of the boat. When discussing the electromagnetic force, we use the term pilot wave instead of bow wave, since the latter tends to be used only when discussing liquids. However, conceptually, both ideas are similar.

As illustrated in Figure 7–14, when particles travel through the primary field they create a pilot wave that travels at velocity w, ahead of the particle, which travels at velocity v. Since the pilot waves propagate away from the particle, they do not represent forces that would alter the particle's direction of travel in the primary field. This same reasoning explains the observations associated with a particle's behavior after passing though a barrier with one slit. However, when two slits exist in the barrier, each serves as wave emission sources in the secondary field. The waves traveling at a faster velocity than the particles form the interference pattern in the secondary field. When the particle crosses the barrier through one of the slits, it enters a secondary field that is not quiescent. The forces resulting from the motion in the secondary field affect the particle's direction of travel. This explanation is consistent with Newton's first law. The particle then finds a peak to ride until it encounters the observation plane, similar to how a surfer finds an optimal place on a wave in the ocean. Technically, we cannot state with certainty that the particle rides the peaks of the interacting waves, however, we can conceptually assume this behavior.

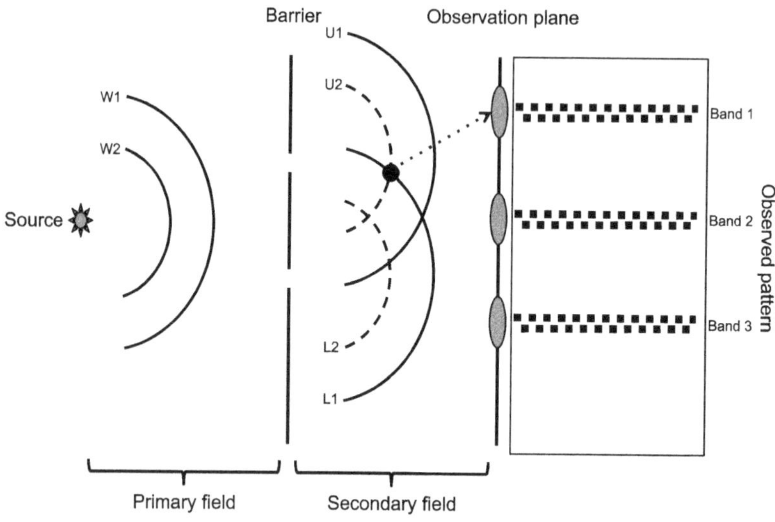

Figure 7–14 Modern Mechanics interaction between waves and particles through a barrier with two slits. In the secondary field, the pilot waves interact and create interference patterns that represent forces, altering the particle's trajectory. In this example, a particle rides the wave produced by the interaction of wave L1 and wave U2.

Because particles will find optimal positions on waves within the secondary field, riding these waves will produce the wavelike interference pattern when the particles hit the observation plane. This will only happen if the following conditions are met. First the pilot wave forms an interference pattern in the secondary field. Second, the particle's velocity must be less than that of the pilot wave. If the particle's velocity exceeds that of the wave, then the wave will not be able to apply any force to the particle in the secondary field to alter its direction. Third, the particle's trajectory has to be influenced by the force of the interfering wave's motion in the secondary field.

This alternative explanation is consistent with Modern Mechanics, classical mechanics, and Newton's laws. It also takes into account the time differences that generate interference patterns. This is a reasonable explanation of the experiment that does not require delayed communication, independent decisions on a particle's direction of travel, or independent changes in a particle's direction that violate Newton's first law.

Louis de Broglie originally proposed the pilot wave theory in 1927. While pilot wave theory was largely dismissed in the mid 20th century, it is currently undergoing a revival. In 2006, two researchers, Yves Couder and Emmanuel Fort, published a paper entitled: *Single–Particle Diffraction and Interference at a Macroscopic Scale*. They showed that pilot wave theory effectively explains behaviors previously associated with quantum mechanics. This type of experiment was not possible at the turn of the 20th century.

In addition, new experiments involving light may one day explain its true nature. Following in the footsteps of Harold Edgerton, who pioneered high–speed photography, researchers at the Massachusetts Institute of Technology are conducting experiments involving light using high–speed photography. They have developed a device that essentially records 1 trillion frames per second. They can observe individual light pulses, which appear to move in slow motion using the technique. This is extremely exciting research because light, something that was once even thought to travel instantaneously, can be observed and studied in slow motion. We can expect that emerging research will heavily influence our understanding of the true nature of light.

Summary

Experiments are an essential part of the scientific process. They serve to provide support to ideas, hypotheses, and theories. They have the power to exclude or eliminate theories as viable alternatives. They also have the power to support a theory or model. However, contrary to popular belief, they do not have the power to prove one theory as the only possible explanation of a experiment or observation.

We must always be mindful of two important characteristics associated with experiments and their relationship with theories. First, multiple theories might be able to explain an experiment. Even in the case where we only know of one theory that explains a theory, we have to remain open to the possibility that one day a new theory might be introduced that explains the experiment with equal or improved accuracy. Second, we have to always remain open to the fact that our analysis of an experiment may change. When this occurs, we might find that an experiment that previously appeared to fail, actually passes. Models that were once excluded may gain renewed interest and attention as reasonable explanations. We might also find that theories based on the experiment's failed result are no longer viable, since the experiment, in reality, passed.

As a unified model, Modern Mechanics must explain a wide variety of experiments and observations. In the ideal case, it explains the same experiments that are explained by classical mechanics, relativity theory, and quantum mechanics. In doing so, it also provides better results and is easier to understand. Modern Mechanics explains many experiments that, when considered together, previously required three models. Its forward and reflected intercept wavelength equations mirror classical mechanic's Doppler equations. Its average intercept equation produces results that improve upon the accuracy of relativity theory's equations. Additionally, these same equations are used to properly interpret the Michelson–Morley interferometer data, revealing that their experiment produced the expected result of $30km/s$. This result fundamentally changes our understanding of space and time, because this experiment supports theories that align with a luminiferous aether.

The success of the experiment has the additional benefit of excluding relativity. These findings are extremely interesting. In fact, not only do these results support Modern Mechanics, they remove support from special relativity theory, as well as from its dependent theories like general relativity and the standard model, since they presume special relativity is correct.

The Modern Mechanics equations provide better results than the relativity equations in the Ives–Stilwell experiment. In addition, we have shown how Einstein's famous energy equation, $E = mc^2$, is actually an approximation. More importantly, we have produced the average intercept kinetic energy equation that is appropriate for a three–system model.

Finally, the double–slit experiment is intuitively explained using Modern Mechanics. In fact, its explanation is consistent with many of the ideas associated with De Broglie's pilot wave theory. Modern Mechanics explains why wave peaks and valleys form at various locations along the observation plane.

The experiments we have examined only scratch the surface of what Modern Mechanics may be able to explain. It opens the door to some interesting research areas and new theoretical possibilities.

Chapter 8 What You Need to Know and What it Means

Modern Mechanics has an ambitious goal: to serve as a unified model in physics. Modern Mechanics introduces ideas and concepts that deepen our understanding of the universe and the world around us. It uses those ideas and concepts to help us improve our understanding of established theories and experiments. Modern Mechanics is unique. It can explain various experiments that previously required three different theories: classical mechanics, relativity theory, and quantum mechanics. This ability to explain how things move with one intuitive, compelling model is extremely powerful.

To serve as a unified model, Modern Mechanics must satisfy several requirements. The first requirement is need: There has to be a reason for the new model. A model that simply provides an alternative explanation will not unseat any established theory, especially one with a century–old track record. Granted, establishing need is not easy. It requires a clear, unequivocal failure in the theoretical, mathematical, conceptual, or observational characteristics of the established theory. This need was initially discussed in Chapter 1, which described the failure of Einstein's spherical wave proof. The mistake was not previously detected because Einstein's proof was incomplete and an examination of the steps he provided would lead one to conclude that the proof passed. Einstein's work was further examined in Chapter 6, revealing specific conceptual and mathematical mistakes in his derivation that render relativity theory wrong. This establishes need.

Need is further established when the new model offers predictions or results that are more accurate than those associated with the historical model. One of the most important experiments associated with modern physics is the Michelson–Morley interferometer experiment. The experiment was designed to detect and measure the change in wavelength of a ray of light while the Earth orbits the Sun. Contrary to a popular misconception, Michelson and Morley did not directly measure the Earth's speed with a speedometer. Instead, they made optical measurements that were used as values in an equation to compute the Earth's speed. They expected their equation would yield an Earth orbital velocity of $30km/s$. What they computed was closer to $8km/s$, which was nowhere close to what they expected. Based on this result, they rejected an aether drag theory that would have explained the behavior of the electromagnetic force.

The adoption of relativity as the leading theory to explain motion has led scientists to interpret the Michelson–Morley result as "experimental error." They argue that the result is actually 0km/s, since this answer aligns with relativity theory. Although the Michelson–Morley experiment was long thought to support Einstein's theory, it does not when the experiment is viewed through a purely mathematical lens. Statistically, a velocity of 0km/s, as required by relativity theory, has less than a 0.1% chance of being correct. More importantly, when the experimental data is analyzed with the proper equations, it reveals that the experiment was a success. They detected $32km/s$ and expected to measure $30km/s$. Interestingly, had the Michelson–Morley experiment been properly analyzed in the 19th century, relativity theory would have never been developed. The success of the Michelson–Morley experiment means the demise of Einstein's theory, since the two are mutually exclusive. Once again, this failure in relativity theory's ability to explain an important experiment involving the electromagnetic force establishes need.

As a second requirement, a unified model must be conceptually and mathematically sound. Modern Mechanics leverages concepts and mathematics associated with geometric transformations, a foundation it shares with classical mechanics. However, Modern Mechanics differs from its historical counterpart in two important ways. First, Modern Mechanics recognizes that scaling and translation equations both apply to inner systems; they simply explain different characteristics. Specifically, translation describes the change in the position of an inner system, while scaling describes distances associated with an inner system's motion. Distance can be unidirectional or bidirectional, the latter of which is an important characteristic of oscillating systems. It is the use of scaling and transformation equations that enables Modern Mechanics to explain observations and experiments that previously required three different theories: classical mechanics, relativity theory, and quantum mechanics.

The second difference is: Modern Mechanics explicitly recognizes the importance of multiple inner systems. This is in contrast with classical mechanics, which only addresses the importance of one inner system. Similarly, Einstein also only addresses the importance of one inner system in his theory. In fact, Einstein's theory confuses behavior and equations by forcing a relationship between the concepts for a nested system relationship and the mathematics for a non–nested system relationship. In contrast, Modern Mechanics addresses the relationship of multiple inner systems, where their relationships are defined as nested or non–nested.

The third requirement is usefulness. A unified model must be more than theoretical; it must improve upon the capabilities and accuracy of the historical theories it replaces. Specifically, Modern Mechanics must explain the same experiments as classical mechanics, relativity theory, and quantum mechanics: It must make predictions equal to or better than the historical theories. Fortunately, Modern Mechanics is an ideal candidate because it explains experiments currently associated with each of the established historical theories. For example, like classical mechanics, Modern Mechanics explains how non–electromagnetic systems move and behave. Like relativity theory was intended to explain, Modern Mechanics explains how electromagnetic systems move and behave. Like quantum mechanics, Modern Mechanics provides an explanation for Young's double–slit experiment.

While not an explicit requirement, a unified model should be intuitive and easy to understand. Since Modern Mechanics is built on the foundation of geometric transformation, many of its core ideas and concepts will appear familiar to many readers. Other ideas or concepts may be familiar, but use terminology appropriate for Modern Mechanics. For example, the Doppler shift equations are extremely important in classical mechanics. They are also used in Modern Mechanics, where they are specific instances of the forward and reflected intercepts. In other cases, Modern Mechanics introduces concepts that may not be familiar to many, but help us view the world in a new way. An example is the explicit use of the subtraction mean equation to find an average. This provided the foundation from which we could reverse engineer Einstein's Tau function to reveal its specific purpose and explain how it works.

While Modern Mechanics shares some similarities with historical models, it is different and unique. It represents a new way of viewing space, time, and motion. Modern Mechanics fully incorporates classical mechanics, negates relativity theory, and could align with the statistical models and equations associated with quantum mechanics. Modern Mechanics makes new assumptions, removing limitations imposed by theories like relativity. It opens the door to new scientific discoveries while providing greater predictive and experimental accuracy.

Philosophically, Modern Mechanics aligns well with Occam's razor. Occam's razor says that *when given multiple choices, the simplest alternative is usually the right one.* Said another way, the model most likely to be right is the one that is: intuitive and easy to understand; conceptually and mathematically sound; provides equal or better answers when explaining experiments and observations; doesn't have mathematical flaws in its derivations or proof; doesn't contain strange paradoxes; and explains observations that previously required three different theories. While well–suited to serve as a unified model, Modern Mechanics' success will ultimately be measured through a historical lens.

What You Need to Know

Modern Mechanics is an intuitive way of explaining motion. It uses discrete types to describe relationships in terms of time and length, and compound types to describe relationships in terms of frequency and wavelength. Modern Mechanics provides a foundation from which to evaluate historical theories and a foundation from which to explore and re–examine important areas of physics. Some of these interesting areas include: time, space, gravity, energy, and faster–than–light communications and travel. Exploring these areas requires answering the question: *What do we know now and where can we go from here?* The first part of this question is answered with the *Top nine essentials you need to know about Modern Mechanics and physics.*

1. Physics helps us explain the world around us.

Physics is an interesting, exciting, powerful discipline. It helps us to answer questions like: *Why does something move? How does it move? What happens when it moves? Where is it after it moves?* Answers to questions like these help us to explain the world around us. Specifically, they help us explain observations and make accurate predictions.

Physics, like computer science, engineering, and other scientific disciplines, relies heavily on mathematics. In many ways, mathematics is no different than a spoken, foreign language. Yet in other ways, mathematics is uniquely different. What makes mathematics special is its degree of precision. It can be extremely exacting and follows specific rules designed to reduce ambiguity. At the same time, it allows us to generalize and develop models with broad applicability. As discussed in Chapter 5, these ideas and concepts are built using a technique called abstraction.

As an example of the power of abstraction, one might use a pronoun in the English language to say: "*She is a college graduate.*" This single sentence can be used to describe many different women. While a complete sentence, the idea only comes alive when we know to whom "she" refers. We do not have to repeatedly write the sentence using every possible woman's name. We just need to define the specific noun that replaces the pronoun. Similarly, physics uses abstraction to create generalized models that we use to explain observations and make predictions.

2. *Classical mechanics is a two–system model that explains changes of position in moving systems.*

Discussed in Chapter 2, classical mechanics is a model consisting of two systems: an outer system and an inner system. The outer system is used as a reference system from which to observe, measure, and explain the motion of the inner system. Motion is measured in absolute terms, with the inner system moving with respect to the stationary system.

Classical mechanics is easy to understand, because it is built upon geometric transformations. It is extremely useful and easily explains many observations and experiments, with one notable exception: It is not able to explain all observations associated with the electromagnetic force. This is a significant shortcoming that prevents it from serving as a unified model. Since the electromagnetic force is the building block of modern electronics and communications, early 20th century physics found itself in crisis without a theory to explain its behavior. This was a crisis that Einstein believes is addressed by and overcome with relativity theory.

3. *Relativity theory is a two–system model that attempted to explain what classical mechanics could not.*

Mathematically, relativity theory explains experiments involving the electromagnetic force. Specifically, it filled a mathematical gap left by classical mechanics' inability to explain experiments like the Michelson–Morley experiment. As discussed in Chapter 6, relativity theory gained its mathematical capabilities from using a scaling transformation instead of a translation transformation. One of the reasons for the success of relativity theory is that scaling equations explain the behaviors of frequency and wavelength better than translation equations. In

Modern Mechanics, scaling equations and translation equations both apply. They just explain different things.

Einstein's scaling–based equations provided useful answers for the Ives–Stilwell experiment and other experiments involving the electromagnetic force. We understand why Einstein's equations often provide acceptable answers: They are normalized versions of the average intercept time and length equations. However, even in cases where Einstein's equations perform well, they will still underperform those associated with Modern Mechanics due to his normalization step.

While Einstein's equations often provide useful results, they suffer from two problems. First, as discussed in chapters 1 and 7, they do not properly describe the results of the Michelson–Morley experiment when they are viewed through a statistical lens. This means they do not adequately explain the electromagnetic force. Second, as discussed in chapters 1 and 6, Einstein's spherical wave proof failed. So, even though Einstein's equations have demonstrated their usefulness, his theory is invalid.

4. Modern Mechanics is a three–system model of motion.

Like classical mechanics, Modern Mechanics is built using geometric transformations. While it shares this very important similarity, it differs from classical mechanics in several ways. First, Modern Mechanics is a three–system model, a characteristic that differentiates it from classical mechanics and relativity theory, which are both two–system models. Throughout this book we have developed equations for a three–system model, however Modern Mechanics will work with any number of inner systems. Second, Modern Mechanics recognizes that translations and scaling equations both apply when describing inner systems; they simply describe different things.

Since Modern Mechanics uses the same translation equations as classical mechanics, both theories make the exact same predictions for observations that do not involve the electromagnetic force. In addition, since Modern Mechanics uses scaling transformations, it can explain behaviors associated with the electromagnetic force, similar to relativity theory. The Modern Mechanics' equations will produce results that will match or exceed the accuracy of Einstein's equations, because of the error he introduces during normalization.

5. *Relativity theory suffers from several problems that cannot be corrected.*

Classical mechanics and relativity theory share a significant similarity: Both are two–system models. However, that is where their similarities end. Unlike classical mechanics, Einstein uses the inner system (not the outer system) as the reference system to observe, measure, and explain motion. This use of relative measurement is where Einstein's theory derives its name.

Although it provides helpful answers, relativity was never a sound theory. However, the fact that it often provides acceptable answers leads to an interesting question: *Can relativity theory be salvaged or corrected?* The short answer is: *No.* Relativity cannot be salvaged, because it suffers from three significant problems:

First, it is a three–system model that Einstein incorrectly describes as a two–system model. Relativity theory is a non–nested, three–system model consisting of an outer system (ie, the stationary system), an inner system (ie, the moving system), and an oscillating system (ie, a ray of light). This is a critical mistake that cannot be corrected while retaining Einstein's theory.

Second, Einstein fails to recognize that the inner system always moves according to the translation equation. Einstein says that the inner system moves according to the translation equation, $x = x' + vt$. He also makes the same statement mathematically as part of the expressions used in his Tau invocations. He cannot repeatedly say that the Newtonian equations are used throughout his derivation and then, at the end of his derivation, conclude that those same equations no longer apply. This mistake cannot be corrected in such a way as to retain relativity as a theory.

Third, Einstein's theory is based on an incorrect treatment and understanding of Tau. Simply stated, the function Tau simply returns the average intercept time. More specifically, Einstein does not understand that Tau is *a function that uses the subtraction mean equation to return the average intercept time (relative to a starting time of t) between the origin and a point (represented by the first three arguments) on the inner system in a non–nested relationship, where the fourth argument is the forward intercept time to that point.*

Tau returns the average time for a ray of light to travel from the origin and a point on an inner system, and from the point back to the origin on the inner system. Called average intercept times in Modern Mechanics, it follows that Einstein's x, y, and z equations are actually the average intercept lengths between the origin and three different points on the inner system. They are not coordinates of a specific point on the inner system. Fundamentally, Einstein's theory is built upon an incorrect understanding of the Tau function and equations. With Tau correctly defined, the behaviors described by the equations can be properly determined. Once again, this mistake cannot be corrected in such a way as to retain relativity as a theory.

Fundamentally, we do not want to salvage a theory that mathematically performs worse than an alternative, is built using an incomplete understanding of the equations, and is conceptually flawed.

6. *Modern Mechanics can explain the quantum mechanics double–slit experiment.*

Quantum mechanics was long thought to be one of a few theories that could explain the Young double–slit experiment. Quantum mechanics explains the behaviors of particles as being similar to that of waves. Unfortunately, this explanation fails to take into account how multiple waves behave to form the pattern observed with the experiment. A proper understanding of the experiment requires us to look at individual waves to see how they form peaks and valleys.

Modern Mechanics explains the observations associated with the double–slit experiment using wave particle *interaction*. Peaks and valleys are formed when waves – *emitted from the barrier at different times* follow paths of different lengths to arrive at various points along the observation plane where they are *observed arriving at the same time.* This explanation describes why particles form peaks and valleys at different points along the observation plane. The Modern Mechanics explanation may be extremely similar to pilot wave theory, which is another non–quantum mechanics theory that explains the experiment.

7. *Einstein's energy equation $E = mc^2$ is an approximation.*

Through the magic of mathematical gymnastics, we can make two different numbers appear equal to one another. Discussed in Chapter 7 as an example, π begins as 3.141 and $\frac{22}{7}$ begins as 3.142. You can visually see that the numbers, while close to one another, differ, starting with the third significant digit. Mathematically:

$$\pi \neq \frac{22}{7}$$

Notice that by truncating both numbers at the second significant digit, one can rewrite these equations as $\pi = 3.140$ and $\frac{22}{7} = 3.140$. Through the use of the transitive relation, one might now conclude that:

$$\pi = \frac{22}{7}$$

This answer is incorrect, since the equations used to arrive at this answer are approximations, not precise numbers. The problem is that the equality operator should not be used with truncated numbers. The truncated statements should be written as $\pi \approx 3.140$ and $\frac{22}{7} \approx 3.140$. At best, we might be able to conclude:

$$\pi \approx \frac{22}{7}$$

but we cannot say that they are equal. Einstein performed this same mathematical manipulation when he developed his well–recognized energy equation. Specifically, he truncated all but the first expression of an important equation called a Taylor series. He then combined that truncated approximation with the kinetic energy equation to ultimately arrive at $E = mc^2$. Since his equation is an approximation, he cannot say that E is equal to mc^2. The best he can say is that E approximates mc^2, or $E \approx mc^2$.

8. *Theories that depend on relativity theory must be significantly revised or completely rejected.*

One of the most important and exciting implications introduced in this book is that many of the theories that make up second–generation physics will need to be revised or rejected. Specifically, the failure of Einstein's relativity theory means that any theory that depends upon its success must now be rejected, or at least significantly revised. This list of theories includes some theories that are quite well–known and widely accepted: general relativity, quantum electrodynamics, the standard model, string theory, and the big bang theory. While it is possible that aspects of these theories might be retained, the retention of any equation, assumption, or conclusion is not automatic. What is more likely is that some of the underlying mathematics from these theories could be revised and aligned with Modern Mechanics, but that the accompanying theoretical explanations will be rejected and redeveloped.

Additionally, the proper analysis of the Michelson–Morley experiment provides support to theories that support an electromagnetic aether. This is an exciting possibility, because the original interpretation of this experiment led to the opposite conclusion. The reanalysis of the Michelson–Morley experiment, using the corrected equations, also paves the way for the re–examination of other experiments long thought to support relativity theory or its dependent theories.

9. Although Einstein's theory is wrong, he still made significant contributions to science.

Einstein was extremely creative and ingenious. It is important to recognize his contribution to science and the difficulty involved in detecting his mistakes. While some may argue for the validity of Einstein's work on emotional grounds or on the basis of a learned belief system, his work can be evaluated mathematically and logically. Einstein's work appears experimentally supported and conceptually sound, making his mistakes extremely elusive and difficult to detect. In fact, the mistakes are not obvious and could not easily be detected without an understanding of namespaces, a critical characteristic that distinguishes functions from equations. In addition, few scientists recognize that the Michelson–Morley experiment has been incorrectly analyzed. The original Michelson–Morley analysis assumed absolute measurements, while the device actually detects relative measurements. These mistakes – one theoretical, the other experimental – are not obvious, or they would have been discovered sooner.

Regardless of the mistakes that invalidate relativity theory, Einstein's work and research has advanced science. In fact, it is the investigation of his work that has led to the research that produced Modern Mechanics. Grounded in this new foundation, we understand how relativity placed limits on just how far we could take physics.

Implications and Research Opportunities

One of the most important characteristics about science is that it does not stand still. Sometimes, it helps us to take steps that further our knowledge along an established path. At other times, it helps us to take giant leaps on a new scientific path. These giant leaps, or paradigm shifts, represent new ways of thinking. Accompanying paradigm shifts are new ideas, concepts, and equations that can significantly alter how we view the world around us, while at the same time paving the way for new discoveries and predictions.

Unconstrained by the limits established by historical models it replaces, Modern Mechanics allows us to make predictions that were not previously possible. One example of a new possibility is faster–than–light travel and communications. This idea is possible because Modern Mechanics does not have a "speed limit." This simple change – the removal of the speed limit that relativity theory demanded – opens the door to ideas and concepts, many of which have yet to be imagined. Some of these concepts and ideas will explain: different propagation mediums, the meaning of time and its implications, and the behavior of gravity.

There is no speed limit.

According to relativity theory, nothing can go faster than light. In fact, there are mathematical problems with Einstein's equations when this specific velocity is met or exceeded. While many scientists treat Einstein's theory as an upper limit on velocity, others have reinterpreted this *speed limit* as meaning nothing can *accelerate* from a velocity less than c to a velocity greater than c. Without explaining how something can initially attain such a velocity, they simply allow the possibility of faster–than–light travel if something is already traveling at that velocity. Regardless of whether a scientist believes that Einstein's theory is an upper limit on velocity or a limit on acceleration, both interpretations suffer from the same problem. They are based on an incorrect understanding of Einstein's equations. Specifically, these interpretations do not recognize that scaling and translation equations always apply and explain different things. They mistreat Einstein's equations as translation equations associated with a two–system model instead of recognizing them as scaling equations associated with a three–system model. Einstein's inner system always moves according to the translation equation, which does not have an upper limit on velocity.

Additionally, both interpretations fail to explain what happens when an object travels at precisely c, where Einstein's equations will always produce an undefined answer. Any theory that addresses the electromagnetic force must explain three cases: where velocity is less than the speed of light, where velocity is at the speed of light, and where velocity is greater than the speed of light. Any theory that does not explain these three cases is incomplete. Fortunately, Modern Mechanics explains all three cases. Additionally, Modern Mechanics is not limited to explaining the electromagnetic force. It can describe different wave mediums, each of which could propagate at a velocity less than or greater than the speed of light in a vacuum. For example, it can be used with water waves, sound waves, and light waves. It might also be used to explain the behavior of yet to be discovered wave mediums with propagation speeds that far exceed *c*. Such wave mediums could explain entanglement, the quantum mechanics phenomenon whereby two distant objects appear to instantaneously interact. A discovery of a *quantum wave medium* would be no less profound that when the Danish astronomer Ole Rømer in 1676 proved that light traveled at a finite speed. We are reminded that prior to his discovery, scientists believed that light traveled "instantaneously."

Faster–than–light communications are no longer the realm of science fiction. For example, L.J. Wang and several of his peers published a paper in 2000 entitled: *Gain–assisted superluminal light propagation.* In their paper, they created specific circumstances where a beam of light appeared to travel faster than *c*. Specifically, a light pulse emerged 62 nanoseconds sooner than it should have if it were traveling at the speed of light. While this may sound like a small value, it was an important experiment because it alone would represent the demise of relativity theory.

While faster–than–light experiments would seem to violate Einstein's theory, researchers go to great lengths to interpret their experiments in ways that remain aligned with it. For example, in support of relativity theory, Wang says: "Information carried by a light pulse cannot be transmitted faster than *c*." However, one could argue that the absence or presence of a signal alone is all that is required to convey information. This is not unlike Morse code, an early form of long–distance communications that is entirely based on the presence or absence of a signal.

With the speed limit lifted, these experiments can be interpreted using Modern Mechanics. New insights into what these experiments might represent could pave the way for new forms of communications, computing devices, and other technologies that have yet to be imagined. In addition to faster–than–light communications, it also opens the door to vehicular travel at superluminal speeds. Although no longer limited by a "theoretical" limit, engineers would still need to determine how to take advantage of these new possibilities.

Time is absolute and relative.

What is time? If you were to look up the definition of *time* in a dictionary, you would find it defined as: *What a clock measures.* This definition simply leads to the question: *What does a clock measure?* The answer to this question is, of course: *Time.* This circular definition is not in the least bit useful. In fact, this circular definition of time, combined with Einstein's mistreatment of time, permeates science and contributes to time being one of the most misunderstood concepts in modern science.

Dr. Glenn Borchardt, author of *The Scientific Worldview*, overcomes this circular reasoning by stating: "Time is motion." He believes that without motion, there is no time, and vice versa. Intuitively, this association of time with motion makes sense, because something must be in motion in order for us to have a time reference. Modern mechanics builds on Borchardt's idea and defines time as *a measurement of motion*. This definition, which is similar to how we define length as a measurement of space, allows us to use time as a discrete type. In Modern Mechanics, time is an abstraction of motion, one that allows us to explicitly use it in theories and mathematical equations.

Modern Mechanics, classical mechanics, and relativity theory all treat time differently. In classical mechanics, time is measured in absolute terms with respect to an outer system. Called *absolute time*, one time reference is elevated above all others. Compare this with relativity theory, where time is measured in relative terms from the perspective of each system. This kind of time reference is called *relative time*. Modern Mechanics recognizes that absolute time and relative time both apply.

Because a Modern Mechanics' model can contain multiple inner systems, any inner system can serve as a time reference. The type of system relationship involved will determine whether absolute or relative time is appropriate. Absolute time references are associated with nested system relationships. Relative time references are associated non–nested system relationships. In Modern Mechanics, *universal time* is defined as a time reference associated with a specific system that is elevated above all other time references. Universal time takes precedence and represents the reference to which other timekeeping systems are calibrated.

Each oscillating system illustrated in Figure 8–1 can serve as a time reference. Systems 5 and 9 measure time in relative terms, since they are associated with non–nested system relationships. The remaining systems measure time in absolute terms, since they are associated with nested system relationships. An oscillating system that moves between two points on the outer system also forms a nested–system relationship.

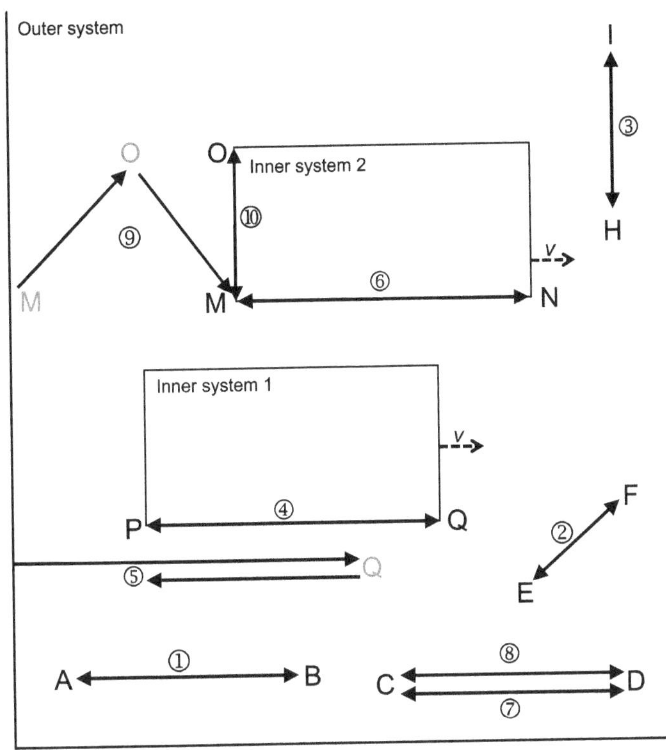

Figure 8–1 Oscillating systems used as time references. Any of the numbered oscillating systems 1 to 10 can be used as a time reference. Some are absolute time references and others are relative time references. In this illustration, systems 5 and 9 are relative references. All others are absolute references.

Ideally, the oscillating system serving as the time reference will be accurate and move back and forth at a constant velocity. However, accuracy and constancy of velocity are not requirements. If the inner system associated with a non–nested relationship is in motion; the length of each oscillation would be greater than if there were no motion. The result is that it takes longer to complete each oscillation, meaning time will appear to run slower. In fact, we can cite several examples where highly accurate timekeeping devices are adjusted to align with less–

accurate references. For example, we use leap seconds, minutes, and days to adjust highly accurate atomic clocks to align with the Earth's solar orbit, which is far less accurate.

Time is extremely interesting and we are now poised to make new discoveries that take advantage of models and theories that build upon universal time, relative time, and absolute time.

Time travel is not possible by going very fast

Time travel is an extremely interesting subject, one that fascinates many in the scientific community. While some ideas about time travel belong in the world of science fiction and fantasy, others find support in the scientific community. Some researchers believe that time travel is supported by relativity theory due to the behavior of its equations once velocity exceeds c. Because Einstein's equations produce a complex number when velocity is greater than c, some interpret this as representing a reversal of time. As discussed previously, this interpretation is based on an incomplete understanding of Einstein's work.

In Modern Mechanics, time travel is not possible by simply traveling faster than the speed of light. Unless oscillations are required in a non–nested system relationship, no constraints are placed on the upper velocity of an inner system. When a non–nested system relationship exceeds the propagation speed of the wave, no oscillations occur because the oscillating system will not be able to traverse the forward segment. This does not limit the velocity of the inner system, nor does it send it hurling backward in time.

Additionally, because Modern Mechanics recognizes the differences between absolute and relative time measurements, all observations are historical. While some things can be perceived as occurring instantaneously, some travel time is always required between the source and observation points. We are always *observing* something that has already happened because it takes time for the light from the event to reach us. Every observation is a glimpse into history, regardless of whether we are watching someone walk down the street or are seeing light from a distant galaxy. The very concept of absolute time, or time measured from the perspective of the outer system, precludes the possibility of going backward in time.

While we have just said that time travel is not possible, the context of the answer is important. Theoretically, one can travel to a position that enables them to see further back in time. For example, imagine that a planet is five light years away from your location and you missed witnessing something important. If you are able to travel faster than the speed of light, you might travel to a point further away – six, seven, or eight light years away from the planet. Since you are able to arrive at that new observation point before the light waves you intend to observe arrive, you can now observe the event you missed. While you are able to see something you missed, which could be considered a form of time travel, you are not actually *traveling* back in time. Instead, you are traveling to a point where the wave you're interested in observing hasn't yet arrived.

Gravity

One extremely interesting idea associated with gravity is called the *push model of gravity,* which was originally proposed by Nicolas Fatio de Duillier in 1690 and again by Georges–Louis Le Sage in 1748. In a push model, we imagine waves pressing against an object from all directions. When a large body gets in the way, it essentially creates a shadow that blocks some of the gravity waves from pressing upon the object from that direction. Since waves that would have pressed against an object are blocked, the waves from the opposite direction will now exert their effect on the object; in effect, they *push* the smaller object toward the larger object. The prevailing gravity model suggests that large bodies possess attraction properties that pull smaller objects toward them. The net effect of either model is the same: the smaller object and the larger object get closer to one another.

A push model of gravity would lend support to the ideas and theories of the late Dr. Thomas van Flandern, who believed in the existence of a gravity medium. This gravity medium is theorized to have propagation characteristics many orders of magnitude faster than *c*. Since Modern Mechanics supports many different types of propagation mediums, it might be used to support the ideas of Van Flandern. The door is open to the possibility of gravity or quantum mediums with propagation velocities much faster than the speed of light.

Conclusion

The goal of Modern Mechanics is to serve as a unified theory of motion. It provides a new scientific foundation: one that is mathematically sound, easy to understand, overcomes limitations associated with existing models, and explains experiments that previously required three different theories. Adopting a new unified model will not occur overnight, The process will follow a path that includes several steps. The first step is awareness. During this phase, people will learn that the new model exists. The second step is education. During this phase people will read the material and learn what Modern Mechanics is and how it might apply to their work. This phase will also bring the most critical reviews and critiques of the model as people re–examine their beliefs in the historical models. After passing through this phase, we will arrive at the third phase, which is called *the tipping point*.

At the tipping point, we will see the rapid adoption of Modern Mechanics. People will begin to use it to explain their experimental findings, while others will use it as the foundation for their theoretical research. It is at this tipping point that a paradigm shift will have occurred. At this point, we accept the new model and will have moved beyond the old model. While the adoption of Modern Mechanics will require passing through these phases, there is no time requirement for how long each phase might last. Of course, the faster we progress through these phases, the faster we will find ourselves on a new path from which we can make new scientific discoveries

Science does not stand still. In fact, this constant exploration and discovery process is one of the most important characteristics that makes science exciting. What we knew yesterday is different from what we will know tomorrow.

Suggested Readings

There are several important mathematical and scientific works that should be considered required reading by anyone interested in understanding physics and motion. The suggested books and papers are collected into three groups: theoretical, experimental, and foundational. The theoretical material focuses on the original works that form the basis of classical mechanics, quantum mechanics, and relativity theory. The experimental material focuses on the original papers that explain key experiments discussed in this book.

The foundational material is a collection of textbooks that covers a wide range of material discussed in this book, including: acoustics and waves, formal mathematical proofs, functions, namespaces, partial differential equations, relations, transformation, and types. This collection does not present all of the material in any given area, nor does it identify which works might be considered the most important. Instead, it presents material that the author found interesting or important.

Another valuable resource is an Internet–based tool called Wolfram Alpha. This is a powerful tool that simplifies modern research and computation, especially when compared with the mathematical effort that founding researchers had to undertake. This level of computing power means that modern researchers can play what–if scenarios to test new ideas in seconds, where in the past, researchers would have spent hours, days or perhaps longer.

Theoretical Material

Albert Einstein, *Relativity – The Special and the General Theory*, 1961

Albert Einstein, "The Foundations of the General Theory of Relativity", in *The Collected Papers of Albert Einstein*, 1997 (146–201)

Albert Einstein, "Zur Elektrodynamik bewegter Körper" (Translated – On the Electrodynamics of Moving Bodies), *Annalen der Physik* 1905 (**17**:891)

George Braziller, *1912 Manuscript on the Special Theory of Relativity*, 2003

Gérard Gouesbet, *Hidden Worlds in Quantum Physics*, 2013

Hendrik Lorentz, Albert Einstein, Hermann Minkowski, and Hermann Weyl, *The Principle of Relativity: A Collection of Original Memoirs on the Special and General Theory of Relativity*, 1952

Henri Poincaré, "On the Dynamics of the Electron", *Rendiconti del Circolo matematico di Palermo,* 1906 (**21**:129)

Issac Newton, *The Principia: The Mathematical Principle of Natural Philosophy*, 1687

Max Planck, "On the Law of Distribution of Energy in the Normal Spectrum", *Annalen der Physik*, 1901 (**4**:553)

Steven Bryant, "Failure of Einstein–Lorentz Spherical Wave Proof", *Proceedings of the NPA Conference*, California State University Long Beach, 2010

Steven Bryant, "The Twin Paradox: Why it is Required by Relativity", *Proceedings of the NPA Conference*, University of Maryland, 2011

Experimental Material

Albert Michelson and Edward Morley, "On the Relative Motion of the Earth and the Luminiferous Ether," *American Journal of Sciences*, 1887 (**34**:333)

Dayton Miller, "The Ether–Drift Experiment and the Determination of the Absolute Motion of the Earth," *Reviews of Modern Physics*, 1933 (**5**:203)

Steven Bryant, "Revisiting the Michelson–Morley Experiment Reveals Earth Orbital Velocity of 30 km/s," *Galilean Electrodynamics*, 2008 (**19**:51)

Thomas Young, *A Course of Lectures on Natural Philosophy and the Mechanical Arts*, 1807 (457)

Thomas Young, "The Bakerian Lecture: On the Theory of Light and Colours," *Philosophical Transactions of the Royal Society of London*, 1802 (**92**:12)

Yves Couder and Emmanuel Fort, "Single–Particle Diffraction and Interference at a Macroscopic Scale," *Physical Review Letters*, 2006 (**97**:154101)

Foundational Material

Alfred Aho, Monica Lam, Jeffery Ullman, and Ravi Sethi, *Compliers, Principles, Techniques & Tools*, 2006

Bjarne Stroustrup, *The C++ Programming Language*, 1997

Earl Swokowski, *Calculus with Analytic Geometry*, 1983

Ethan Bloch, *Proofs and Fundamentals – A First Course in Abstract Mathematics*, 2011

F. Alton Everest and Ken Pohlmann, *Master Handbook of Acoustics*, 2009

James Stewart, *Calculus – Early Transcendentals*, 2008

Jiri Matousek and Jaroslav Nesetril, *An Invitation to Discrete Mathematics*, 2008

John Hughes et al, *Computer Graphics Principle and Practice*, 2014

Raymond Serway and John Jewett, *Physics for Scientists and Engineers with Modern Physics*, 2004

Richard Feynman et al, *Lectures on Physics*, Volume 1, 1963

Tools

Wolfram Alpha, www.WolframAlpha.com, (website current at time of publishing)

Index

A

F

N

O

P

Q

R

S

T

U

V

W

Y

www.ingramcontent.com/pod-product-compliance
Lightning Source LLC
Chambersburg PA
CBHW061624220326
41598CB00026BA/3873
9780996240901